Tragedy at Second Narrows

An indispensable link in the Lower Mainland's transportation infrastructure, the Ironworkers Memorial Second Narrows Crossing, fifty years after construction.
PHOTO ERIC JAMIESON

TRAGEDY
at Second Narrows

The Story of the Ironworkers Memorial Bridge

Eric Jamieson

HARBOUR PUBLISHING

2 3 4 5 — 12 11 10 09

Harbour Publishing Co. Ltd.
P.O. Box 219
Madeira Park, BC V0N 2H0
www.harbourpublishing.com

All photographs courtesy of the author unless otherwise stated
Re-creation of causeway illustration on page 50 by Joan Jamieson
Cover design by Anna Comfort
Printed in Canada
Printed on chlorine-free, FSC certified paper made with 30% post-consumer fibre.

Harbour Publishing acknowledges financial support from the Government of Canada through the Book Publishing Industry Development Program and the Canada Council for the Arts, and from the Province of British Columbia through the BC Arts Council and the Book Publishing Tax Credit.

THE CANADA COUNCIL | LE CONSEIL DES ARTS
FOR THE ARTS | DU CANADA
SINCE 1957 | DEPUIS 1957

BRITISH
COLUMBIA
ARTS COUNCIL
Supported by the Province of British Columbia

Library and Archives Canada Cataloguing in Publication

Jamieson, Eric, 1949–
    Tragedy at Second Narrows : the story of the Ironworkers Memorial Bridge / Eric Jamieson.

Includes bibliographical references.
ISBN 978-1-55017-451-9

    1. Second Narrows Bridge Collapse, Vancouver, B.C., 1958. 2. Bridge failures—British Columbia—Vancouver—History. I. Title.
TG27.V35J34 2008   363.11'962420971133   C2008-904970-5

This book is dedicated to the memory of those who lost their lives in the process of constructing the Second Narrows Bridge. They are as follows:

Albert Bearchell

Sydney A. Belliveau

Walter H. Carter

Joseph Chrusch

Kevin Duggan

Stanley M. Gartley

Alexander Hauga

Frank Hicklenton

Rudolph Hoelzl

Loen Joyal

Gordon MacLean

Alan MacPherson

J.A. Murray McDonald

John B. McKibbin

Richard L. Mayo

Percy Moffatt

Leonard K. Mott

Alexander Robertson

Roderick L. Smith

Alan C. Stewart

John Thompson

Thomas A. Warswick

John A. Wright

Once you have tasted work on a bridge, it is hard to go back to anything else. "Bridgework gets into your blood," reflects Gary Poirier, ironworker, "and it never lets you go." PHOTO P. STANNARD, DOMINION BRIDGE CO. LTD.

Contents

The two cantilevered arms of the new Second Narrows Bridge meet in the middle of the Inlet, drawing to a close one of Canada's most tragic bridge construction projects. OTTO LANDAUER OF LEONARD FRANK PHOTOS, JEWISH MUSEUM ARCHIVES LF-37081

A signalman acts as the eyes of the operating engineer manipulating the controls of the giant traveller as it booms down a vertical and gusset assembly to the raising gang. PHOTO P. STANNARD, DOMINION BRIDGE CO. LTD.

Foreword

It is the nightmare of every bridge engineer that he or she will make a mistake that will result in the collapse of a structure, and worse, loss of life. Being mere mortals, we can make mistakes. Realizing this, engineers strive to catch any errors that might otherwise slip through our nets. Mostly, we are successful. Sometimes—fortunately rarely—we are not.

Although we are all fallible—engineers, ironworkers, inspectors, all of us—and mistakes sometimes happen, it is worth remembering that bridges are among the safest man-made structures on earth. There is no need to worry as you cross over the next bridge. But do drive carefully, because driving is thousands of times more dangerous than being on a bridge.

Having been a bridge engineer for nearly half a century, I wondered what I could possibly learn from a book such as this. I will tell you. Author Eric Jamieson has brought alive the events and drama that led up to, surrounded, and followed the Second Narrows Bridge disaster. When he gave me the manuscript to read, I found that I devoured it. I was captivated by the way he describes the hopes, fears, sadness and courage of the people affected by the event.

I arrived in Vancouver after the bridge was rebuilt, and since then I have worked with some of the people involved in the tragedy, which strengthened my interest. One winter I helped build a bridge in Castlegar with Gary Poirier, an ironworker who had gone down with the bridge. Our foreman was Roy Clemenshaw, who had assisted in the rescue. I developed enormous admiration for Don

Jamieson and Bob Harris, two of the engineers who rebuilt the bridge and who went on to have very distinguished careers in the bridge-building business.

Finally, a tribute to all who design and construct bridges: I have had a wonderful career as a bridge engineer, and have enjoyed the dedication and friendliness of the men and women with whom I have worked. I have learned much from them. I am proud to be a bridge engineer in their company and proud that my daughter is engaged to marry an ironworker.

–Peter G. Buckland, C.M., P.Eng.

Preface

June 17, 1958, began for most of Vancouver as a warm spring day, almost predictable given the imminent summer solstice and the city's location on the steamy west coast of British Columbia. The forecast called for mainly clear skies until early afternoon when cloud cover would predominate. The temperature was to climb to a balmy 25.6 degrees Celsius by mid-afternoon, with a mild breeze from the west. I was to be nine years old the following month, and little mattered to me but that the day was warm, sunny and full of promise. Summer holidays and freedom were but a breath away, and the only niggling anxiety in my young mind was what my grade-three teacher, Mrs. Read, would write in my report card.

Frankly, from my standpoint, the day was unremarkable except for what would occur later that afternoon, a tragedy that would be indelibly burned into my conscience. Everyone recalls pivotal moments in their personal lives, events and memories associated with them that serve to define, for both ourselves and others, who we are. And then there are those other events, external to our personal associations, that tend to invade the public consciousness. We may not know much about the event, but we sure know what we were doing at the time of its occurrence. The following tragedy was such an event for me and for thousands of others living in Vancouver at the time.

Late in the afternoon of that muggy Tuesday, the waters of Burrard Inlet's Second Narrows exploded open with the impact of hundreds of tons of fabricated steel that only seconds before had been a partially erected bridge cantilevered out over the flooding tide. Fourteen ironworkers, two engineers, an operating engineer and a painter were crushed to death or drowned, and a few days later the rescue effort claimed the life of a diver. The event was broadcast to a shocked public through the dramatic voices and footage of onsite radio and TV reporters, and later by the solemn words and pictures of the print media. Many in Vancouver had never witnessed a tragedy of this magnitude.

Prior to 1958, only three catastrophes in the Vancouver area came close to comparison, the first being the great fire of June 13, 1886, where eight people died. Roughly thirty-nine years later, on February 7, 1925, eleven Japanese crewmembers drowned after their ship's tender ran afoul of a barge, and twenty years after that, on March 6, 1945, eight longshoremen were killed in the *Green Hill Park* explosion.

Despite the passage of fifty years and the building of a global metropolis, the Second Narrows Bridge disaster still stands as the worst industrial accident the city has ever experienced. The question in my mind all these years is still, How could it possibly have happened? This, and the stories of the men who built it, lived through the collapse and reconstructed it, led me to tell this story. Men, as vulnerable in body as any of us, but in courage, resolve and spirit, as tough as they come.

The story of the collapse has also become somewhat of an urban legend in Vancouver, and as legends do, they tend to distort with each telling. I recognize that these stories are important, as sometimes they are the only bit of history one hears, but just as often they are so far removed from the truth they are ridiculous. I hope that in this telling I have not added to their credibility, and perhaps the confusion over what really occurred.

Once I had started my research, it was a series of synchronicities that convinced me that I was on the right path, but these serendipitous encounters left me scratching my head with the surreality of them all. The most startling was that a few months after beginning, my son introduced my wife and me to his sweetheart—now his wife. As it happens, her grand-uncle was Alan Stewart, an ironworker apprentice who died in the collapse and whose body was never recovered. So for Alan, and the rest of the men who died and those who survived Vancouver's worst industrial accident, this is their tale.

NB: It is out of no disrespect that I refer to the bridge as the Second Narrows Bridge. I do this out of a sense of history only. The bridge was rightly renamed the Ironworkers Memorial Second Narrows Crossing in 1994 to remember the men who died.

On the afternoon of June 17, 1958, eighteen men were crushed or drowned when Vancouver's partially constructed Second Narrows Bridge plummeted into the waters of Burrard Inlet. A few days later the rescue effort claimed the life of a diver, as a solemn Vancouver watched on. In 1994 the bridge was renamed The Ironworkers Memorial Second Narrows Crossing in honour of the men who lost their lives during that tragic event. *THE PROVINCE*

The first piece of steel on the south shore is lowered into position over the anchor assembly of Pier 17. PHOTO P. STANNARD, DOMINION BRIDGE CO. LTD.

1

The Fifties: A Time of Change

Eighteen men just as hard as nails
And as eager as a brand-new bride
And as the clock on the tower said quarter-of-four
All eighteen had died

—FROM "STEEL MEN"
(WORDS BY DAVID MARTINS, SUNG BY JIMMY DEAN)

A sunny morning fishing on a small lake in British Columbia's southern interior, mid-July 1958, was just what the doctor ordered. Jim English, job superintendent for Dominion Bridge, was alive, and ever so thankful that he was able to share more time with his young wife Ruby and his two children: son Pat and daughter Lynda. English had just witnessed eighteen of his colleagues die in a horrific industrial accident, and he was still bathing in the relief of having been spared the same fate. English was injured, but not so badly that he couldn't pull the oars of the small clinker-built boat through the sun-shot green waters of the lake. He was damn well going to enjoy his recuperation.

As is usual with children and trout fishing, their rods needed constant attention: lines to untangle, weeds to remove and worms to replace. It was a steady shuffle between rowing and rod mending. English also had a rod out, which was firmly lodged between his feet. Occasionally, when he had to stop to tend to his children's needs, he would pass it to Ruby to hang on to. Ruby was not so much interested in fishing as she was just being with her family. In fact, she didn't even own a fishing licence.

Everyone in the family heard the motor of Allan Gill's boat approach, but gave it little notice; the lake was full of motorboats. Although Gill was the local conservation officer, English and his family were doing nothing illegal, so what could possibly go wrong on such a pleasant day?

"Morning folks," Gill said. "Your licence, Ma'am?" he asked.

"I don't have one," Ruby replied. "I'm just holding my husband's rod while he fixes up the kids."

"I've heard that one before," Gill said as he reached over and grabbed the rod from Ruby's hand. "There's a fine for this, you know, and I'll have to confiscate this rod," he said, sternly.

"Is there jail time?" Jim asked, smirking mischievously.

"Well, there can be if she doesn't pay the fine," Gill responded, seriously.

"Well kids, can we do without mom for a couple of weeks?" Jim asked, turning to his children who quickly burst into tears.

"Some holiday this is going to be," Ruby said angrily. "First, my husband falls off the Second Narrows Bridge and barely escapes with his life, and now this."

"Is that true, sir?" Gill asked.

"Yeah, it is," Jim replied, slowly.

"Well, folks, have a nice day, but get off the lake," Allan said respectfully as he passed the rod back and tore up the ticket.

It was the end of the fishing trip for English and his family, but not the end of the recognition received by him and close to two dozen others who had ridden the collapsing spans of the partially completed Second Narrows Bridge into the churning waters of Vancouver's Burrard Inlet and lived to tell the tale. That they survived made them larger than life; that they had done it with such aplomb, made them heroes. But these were the post-war fifties and people were ready to embrace new marks of distinction, ones that neatly fit the new cultural, social and economic realities that were unfolding. English and his gang easily fit this bill.

The energy during that era was palpable, manifesting itself in new forms of cultural expression, particularly music. It was a time of magic: a slow dance at the sock hop with the glint of chrome on cherry hotrods rumbling just under the dulcet tones of the Platters. It was the century's sigh of relief: the Second World War had ground to a grisly halt the previous decade and except for the conflict in Korea, the continent was busy getting back to the freedoms that the war had so valiantly won for all of us. The boom times were here.

The ever-savvy media were tuned in as well. The fifties began with people reading newspapers and listening to radios—and ended with them watching television. News was fast becoming visual entertainment and in 1958 we were all invited to witness firsthand a cavalcade of new and enlightening characters and events: the bridge collapse and the province's one hundredth birthday vying for airtime. It is safe to say that right across the nation it was a tumultuous time. In the province of British Columbia, life was mostly calmer, though we had our own sets of trials, tribulations and firsts.

Most notably, on August 1, 1952, W.A.C. Bennett became our twenty-fifth premier. He would soon start his campaign of road- and bridge-building with his effusive Minister of Public Works (later Minister of Highways), the infamous Flyin' Phil Gaglardi. Falling neatly into the government's aggressive infrastructure agenda, the present-day, freeway-style Granville Street Bridge was built and funded by the City of Vancouver, opening with great fanfare on February 4, 1954. The next year, the Province purchased the Lions Gate Bridge, constructed in 1938, for $5,959,060 (well below its $10,032,540 appraised value). Later in 1955, construction began on the new Second Narrows Bridge. In the summer of 1957, the Oak Street Bridge opened to traffic.

But bridges weren't all that occupied the province during that era, and the year 1958 was probably the most diverse and memorable of the decade. It was equal parts ceremony, drama, excitement and tragedy, and sometimes all four wrapped up into one. The province was busy preparing to celebrate its centenary, having officially become the Crown Colony of British Columbia by an act of the British Parliament on August 2, 1858.

One hundred years later, Bennett's government was preparing to celebrate a century of progress. Included in the celebrations was a visit from Her Royal Highness Princess Margaret, who was to attend some of the official ceremonies. A consular corps ball, a festival of the arts, a jazz festival and an international ski meet were organized, along with a centennial games including champion mile runners, and the largest golf tournament ever held in Canada. Topping it all off was the announcement that a new maritime museum was planned for Vancouver as a permanent centennial memorial. The museum was to house the iconic RCMP schooner, the *St. Roch*.

An interview Bennett gave at the time was revealing. Ever the prognosticator, he forecast that within the next hundred years there would be a staggering surge in the province's population—from approximately 1.5 million in 1958 to 25 million by 2058—with megacities dotting the province. Perhaps that is what motivated him to adopt one of the most impressive highway and bridge construction campaigns ever mounted in the province by any government before or since; a level of development that earned his strategy the nickname "blacktop government." The Second Narrows Bridge figured prominently in those plans.

The drama of that year is evident in a yellowed *Province* newspaper clipping dated July 2, 1958, that my paternal grandfather, Edgar A. Jamieson, P. Eng.,

With Public Works Minister Phil Gaglardi looking on, Premier W.A.C. Bennett turns the sod of the new Agassiz-Rosedale Bridge, thus kick-starting a campaign of road and bridge building unparalleled in the western world. BC ARCHIVES I-68817

had inserted in his diary. It read: ROADBLOCKS, FLOODLIGHTS, GUARDS—CAN THIS BE CANADA, 1958? The report, from Nelson, detailed police roadblocks scattered throughout the Grand Forks, Rossland–Trail, Nelson and Slocan Valley areas in the RCMP's relentless search for the perpetrators of several recent bombings. Everyone blamed the Sons of Freedom and Orthodox Doukhobors, whose radical sects occasionally resorted to violence to register their protest against such things as mandatory schooling. The word "terrorism" was used in the first paragraph of that article, a word that now fills our newspapers on a daily basis, but was then uncommon.

The excitement of that year, to a young boy anyway, was an event that I will never forget. On April 5, 1958, I sat in the basement of my grandparents' home just outside of Victoria and watched on their black-and-white television set as the CBC aired one of its first nationwide live broadcasts, the annihilation of the infamous Ripple Rock. The treacherous, twin-peaked underwater reef, which loomed menacingly close to the surface of Seymour Narrows, had already claimed 114 vessels and 125 lives.

But this event did not stir the public consciousness more than the next, an industrial accident unparalleled in the history of the city of Vancouver. Named by the *Vancouver Sun* as the most important story of the year, the collapse of the partially completed Second Narrows Bridge, and the subsequent loss of life, took the city by surprise. Although this would not be the only bridge tragedy in BC history, it would rock the province as though it were. On another balmy Tuesday, almost sixty-two years to the day earlier, the Point Ellice Bridge across Victoria's Harbour collapsed, killing fifty-five men, women and children on their way to a May Day celebration.

But the Point Ellice Bridge was an old wooden structure that had been poorly maintained while the Second Narrows Bridge was a modern steel bridge under construction. A vital link in the province's chain of new highways development, the bridge was perhaps a symbol of the government's intent. The collapse not only tarnished the Highways department but only four days after the collapse, brought down a rain of abuse on the Highways minister's head. Opposition leader Robert Strachan left no stone unturned in his attack.

Of course, both the minister and his ministry were easy political pickings. Not only was Gaglardi's bombast a frequent target, but the Highways ministry's annual expenditures was a favourite also. Prior to the 1952 election that put Bennett into power, less than 10 percent of the provincial budget had been allocated to road development, but by the end of the decade, with Gaglardi at the helm, it had soared to over 20 percent. Mel Rothenburger documents this transformation in his biography *Friend o' Mine: The Story of Flyin' Phil Gaglardi*: "Between 1952 and 1958, Gaglardi built 825 miles of new highway under contract, surfaced or resurfaced 2,800 miles and rebuilt or improved another 2,900 miles with his own crews.... [I]n the late 1950s and early 1960s, Gaglardi finished the Rogers Pass, the Vancouver-to-Washington and Vancouver-to-Hope freeways, the Upper Levels highway and Burnaby's famous sawdust highway. When he wasn't building brand-new highways, he was rebuilding old ones—the Hope-Princeton, the Fraser Canyon, the Southern Trans-Provincial, Highway 97 through the Okanagan and Cariboo. They were tied together with spectacular bridges like the Second Narrows, the Port Mann, the Alexandra, the Agassiz-Rosedale and the Princess Margaret."[1]

Phil Gaglardi had more monikers, whether out of affection or derision, than any other politician in my lifetime. He invented the famous *Sorry for Any Inconvenience* signs that frequented his many highway projects, and thereby became known as "Sorry Phil." Bennett called him "the greatest Roman road builder of them all," and he was later known as "Flyin' Phil," not only because of his penchant for collecting speeding tickets (of course he was just testing the highways for safety), but for his proclivity to fly. He persuaded Bennett to purchase a Lear jet and that jet eventually landed him in deep trouble when he was alleged to have arranged its use for his relatives.

He worked tirelessly and when he wasn't attending to road construction he was pursuing his other love as pastor of the Calvary Temple Pentecostal Church at Kamloops, a position he maintained even while in government. When he did have a spare moment for his growing family, they would spend it together in a car inspecting various highway projects around the province. Despite his tight agenda and general state of activity, Gaglardi was deeply affected by the bridge collapse. "I know that the two things that I remember discouraged him, hurt him very badly, was that [the bridge collapse] and that landslide . . . that happened at Hope," says his son Bob. "His whole life was about helping . . . helping people, helping them through hardships . . . and that was what my father was about . . ." Although the government was exonerated from blame by Sherwood Lett, the head of the royal commission called to investigate the collapse, Gaglardi had his pastor's hat on when he remembered the lost men and their widowed families.

The disaster also brought a measure of reflection to most of the citizens of Vancouver, settling into the public consciousness like a bad dream. Although half a century has passed, most people remember what they were doing on that "Black Tuesday" with startling clarity. I was just coming home from school late that afternoon, and as I came loudly into the kitchen my mother shushed me as she listened intently to our little Westinghouse radio squawking in the corner of the counter. CKWX announcer Red Robinson broke the melody of "Do You Wanna Dance?" by Bobby Freeman with the news that brought the city to a standstill: "Well, for the past two hours you've been listening to the afternoon show. Stay right where you are on CKWX for further reports on the Second Narrows Bridge tragedy. Keep your ear right here on CKWX Radio. This is CKWX Radio British Columbia . . ."

Reporter: " . . . saying very little in that the ambulance service is still coming and going, possibly a little more frequently, but the thing that we're deploring here, along with the police, is the number of pedestrians that are here on the scene. The list of casualties is unbelievable and the police say these police are going to be here probably for one, a very long time . . . the divers have been in the water here . . . it will take a long, long time to shift the load, and if people want to come sightseeing, the police advise to stay away from the area, particularly either this week and next, but don't come tonight. The bridge is closed to all traffic . . . they're keeping it open for ambulances and other emergency vehicles, and definitely closed to all other traffic . . ."

Red got the scoop of the decade largely because of his paternal uncle Chuck, who had been relaxing on his sister's front lawn sipping a beer with his brother-in-law when the bridge collapsed. From the house, located on Wall Street in Burnaby just near Burrard Inlet, they had a decent view of the bridge. Coincidentally, they had been discussing it, and what they considered to be its flimsy supports, when they heard a loud report.

"We turned back again and everything was . . . it was like it was in slow motion, but we could see some of the workers going down with the crane and everything and boy there was a lot of noise and commotion going on there . . . but as soon as it happened I said to my sister; 'Hand me the phone out the window, I'm going to phone Red.' And when I phoned, I said, 'I'd like to speak to Red Robinson right away,' and she said, 'I'm sorry, he's on the radio,' and I said, 'Well, this is Uncle Chuck and the new Second Narrows Bridge just fell down,' and she put me right through to Red and he started asking what was going on."

"One of the people it hit real hard too was Phil Gaglardi," Chuck recalled. "Al Oster, at that time was the Superintendent of Traffic, and Al and I were real good friends, and I went over to Al's this one night and Phil Gaglardi was there . . . and we're sitting having coffee at Al's place and then Al's wife, Alice, turned around and gave Phil a hanky and we were talking about the bridge. So it hit him pretty hard I tell you."

Even before Red's shift had ended that day, the question on everyone's mind was clear: what had gone wrong—bridges just don't fall down, do they?

2

The Crossing Argument

*Some people want to sit around arguing and haggling over
where it's going to be. We can do that after the bridge is built.*
–PHIL GAGLARDI, MINISTER OF HIGHWAYS[2]

The age of prosperity had come, and the economic growth that accompanied
it created an urgent need for more homes to accommodate the wealthier and
expanding population. The North Shore had become a popular bedroom community to the growing city of Vancouver due to cheaper land and taxes, and subdivision after subdivision were being carved from the slopes of the North Shore
Mountains. Property was still relatively inexpensive by today's inflated standards,
but far more than the $300 lot price advertised by S.F. Munson, real estate broker,
in 1911. Then, $20 cash and $15 per month would purchase a view lot overlook-
ing the Second Narrows. By 1958, prices had risen to the princely sum of $2,950
for the same lot and housing in the area anywhere from $10,000 to $17,000.

But prosperity was a double-edged sword. While the rate of growth was eco-
nomically healthy, it also exposed a critical weakness. The existing bridges and
ferry system across Burrard Inlet were inadequate to meet current needs, a condition that had already been identified by the mayor of the City of North Vancouver, Frank Goldsworthy, as early as the fall of 1951 when he stated: "With the increase in traffic over the narrows, we should have a more modern span, equipped
to accommodate a greater flow of traffic."[3]

The increase in traffic was due to a number of factors, but none so preva-
lent as the increasing reliance on the automobile. It was during the fifties that
the automobile burst from its cocoon, morphing from drab pupae to winged
wonder. Models were being manufactured that are still being idolized such as the

Known as the "Bridge of Sighs" because of the number of marine accidents, the antiquated structure had up to sixteen interruptions a day from both marine and rail traffic. LEONARD FRANK COLLECTION, VANCOUVER PUBLIC LIBRARY VPL-12379

'57 Chevy Bel Air, a stock dream car synonymous with muscle, music and sex. Automobiles of the fifties were suddenly more than just transportation; they were symbols of status and success. The era was known as the "Golden Age of Automobiles," and although fifties' marketing efforts had much more to do with image than reality, they worked nevertheless. Gone were the humpbacked coupes of the forties and here were a dizzying array of colours, mile-long fins, sharky grills, burnished chrome and wraparound windshields that went on forever.

The revolutionized automobile, together with prosperity and urbanization trends, burst the seams of the city's existing crossing infrastructure. The old two-lane Second Narrows rail-and-car bridge, constructed in 1925, crossed the Second Narrows just east of where the current bridge is now. The crossing system was completed by the three-lane Lions Gate Bridge, and a car ferry service between North Vancouver and Vancouver. In 1953, two ferries moved 1.3 million people and 155,000 vehicles across the harbour, no small feat considering that the maximum payload for each vessel was only thirty vehicles per trip. Harland Bartholomew, a respected Vancouver planning consultant, had envisioned a fast ferry system linking various coastal locales with the city. But his idea never caught on, likely because the emerging dependence on and relative freedom of the automobile made it redundant.

Of the three crossing options, the ferry system was slow and problematic. But even the ferries couldn't match the frequent delays that plagued the old Second Narrows Bridge. It had a lift span (originally built with a bascule span but replaced by a lift span in 1934) that would rise or lower according to the whims of marine traffic. Cars would travel on either side of the iron-grated bridge deck while the CNR tracks ran through the middle. Traffic interruptions were legendary, and in 1952, bridge tenders recorded 28,381 boats passing under the structure, requiring the bascule span to be raised 5,315 times. In addition, almost seven hundred trains crossed the bridge that year for a combined average of about sixteen interruptions a day. Although much of the marine traffic didn't require the complete lift of the span, it was frustrating nevertheless.

Raising the span was a complex affair that required quick but calm action once the captain of a ship wanting passage whistled the bridge tenders sitting in the control room thirty-six metres above the water. Warning bells would be sounded, gates dropped in front of motorists and locks released on the moveable span. And all the while the freighter would be steaming toward the bridge at a seemingly reckless pace. The number of marine accidents attested to the danger of the bridge. As bridge tender Al Engleman observed in 1953, "It's no place for a nervous man."[4] If it snowed, the snow had to be cleared immediately because its weight could overburden the counterweight, thus preventing the 816-tonne lift span from opening on demand. This, plus the fact the old structure was gaining between eight and ten tonnes per year in unwanted water weight, spelled trouble.

The Second Narrows Bridge was owned by the Burrard Inlet Tunnel and Bridge Company, a civic body whose shareholders were the City and District of North Vancouver, the District of West Vancouver and the City of Vancouver. City of North Vancouver Mayor Frank Goldsworthy served as president of the company, and was anxious to have the bridge replaced—but the decision would soon be out of his hands. The previous fall, the company had begun the process of transferring the bridge to the Canadian National Railway. Although the agreement with the CNR would not be ratified by public plebiscite for another two years, its main conditions would remain intact: the bridge company would transfer the bridge on condition that the CNR assume the National Harbour Board debt of approximately $98,000, and agree to maintain, rebuild or replace the bridge, if necessary, for the next twenty years. At the end of that period, the structure would be offered to the railway company for the paltry sum of one dollar.

It was unlikely that Mayor Goldsworthy would see a new bridge under this agreement. But something had to be done, and on March 16, 1953, a meeting was held between the mayors and reeves of the City of Vancouver, the City and District of North Vancouver, and the District of West Vancouver; and representatives of the Burrard Inlet Tunnel and Bridge Company and the First Narrows Bridge Company (a Guinness-family corporation that owned the Lions Gate Bridge). The parties decided to form a special committee to investigate the inlet's future crossing requirements. To do the technical work, a subcommittee was struck, composed of the municipal and bridge companies' engineers. They would investigate future crossing needs and recommend the best location for the new structure, the most obvious being at either the First or Second Narrows.

A year later, before the report was issued, representatives of four of the seven east-end North Vancouver ratepayers associations—Deep Cove, Dollarton, Seymour and Seymour Heights—met informally to discuss how they could speed up the bridge project. Hal Denton, spokesman for the Seymour Ratepayers Association, complained that "those of us living in the areas most affected feel a start should be made right away on it . . . the traffic situation right now is beyond a joke, and with two housing developments expected to bring 1,600 new families into this area within the next two or three years, we think it's about time someone started the ball rolling on this."[5] Not only were the ratepayers anxious, but Councillor George Sargeant reflected the attitude of the municipal governments when he demanded that land be purchased immediately to facilitate access to the proposed bridge.

Meanwhile, activity was heating up at the west end of the shore. The First Narrows Bridge Company had a fifty-year franchise, which would expire in 1988, licensing it to operate the Lions Gate Bridge. That agreement not only gave it the authority to charge tolls for the duration of that period, but also obligated it to provide an adequate level of service. Although the service had become tarnished with daily traffic delays, according to lawyers at Vancouver City Hall the company was not legally bound to build another bridge across the First Narrows. Anyway, city engineers were not sure how a new First Narrows bridge would work with the existing one and where precisely it would land in Stanley Park. Rumours that this would be the crossing of choice were rife nevertheless. This speculation was fuelled by sources close to the crossing committee's investigating team, who advised that preliminary plans for a new six-lane $20-million dollar suspension bridge just east of the existing structure were already drawn. Perry D. Willoughby, First Narrows Bridge Company manager, refused to confirm or deny the story.

It was little wonder then that the public was angry. Nobody seemed to be taking responsibility for the commuter chaos. Historically, the Province was responsible for roads leading to and from Vancouver but not linking Vancouver to its sister municipalities—that was a municipal responsibility. And until the connecting highway was built on the North Shore, bridges crossing from Vancouver to the North Shore were considered to be inter-municipal connectors. Once the highway was in place, the inlet bridges were expected to become part of the provincial highway system and thereafter the responsibility of the Province. In early Vancouver planning days, city officials conducted no serious arterial planning, given that there was little need. The need was now very apparent, but the municipal resources for such an undertaking were not.

In the free-for-all environment that was spawned through inaction, in September 1954 the North Vancouver City Town Planning Commission (TPC) entered the fray by announcing their preference for a third crossing. It wasn't at either the First or Second Narrows, but somewhere in between. The 1,800-metre structure they envisioned would launch itself from the North Shore well up the hill at Queensbury Avenue and land in Vancouver at the rise on Nanaimo Street. Their rationale was that pressure would be alleviated on the Lions Gate, and truck traffic—forecasted to increase once the new highway was complete and the eastern terminus of the Pacific Great Eastern Railway (PGE) was built on the North Shore—would be directed away from the downtown core. It all made perfect sense to the TPC, but it was a non-starter nevertheless. North Vancouver District engineer A.H. Ashworth commented that the span would be too expensive due to its length.[6]

Timing is everything and everything is timing as the saying goes, and it was perhaps no surprise that on November 13, 1954, the First Narrows Bridge Company finally put some substance behind the rumour of a new bridge by announcing that it was prepared to entertain construction of a four-lane suspension bridge over the First Narrows. After haggling with the Squamish Nation for the required land, a 4.4-hectare parcel they picked up for a seemingly low $1,875 per hectare, they were ready and willing to proceed. Only a day earlier, the Burrard Inlet Crossing Report Committee, of which the First Narrows Bridge Company was a member, had released its report.

The report, completed more than a year and a half after being commissioned, revealed some interesting yet troubling statistics. Using 1939 as its baseline date, it forecast to 1976 to ensure that whatever crossing was chosen would be adequate in twenty years (population forecasts beyond that were considered too conjectural). In 1939, 7.5 million people and 2 million vehicles crossed Burrard Inlet. By 1953, fourteen years later, crossings had increased to 22.7 million people and 10.7 million vehicles, representing approximate 203 and 435 percent increases respectively.

Traffic congestion on the North Shore was at this point on June 15, 1954, when the 9,000-tonne Norwegian luxury freighter/passenger vessel *Bonanza* found herself in trouble on a flood tide. She broadsided the old bridge, skewering

On September 19, 1930, the log barge, *Pacific Gatherer*, under tow by the tug, *Lorne*, ran afoul of the old bridge, toppling the centre span into the Inlet, which eventually caused the bankruptcy of the North Shore municipalities. The bridge was not rebuilt until 1934. LEONARD FRANK COLLECTION, VANCOUVER PUBLIC LIBRARY VPL-3115

her bow to the structure. It not only shut down the bridge, but also caused such a traffic snarl at the Lions Gate that police appealed to motorists to leave their cars at home. The lineup for the North Vancouver ferry extended blocks up Lonsdale Avenue as exasperated commuters cursed their plight. The old bridge, derisively known as the "Bridge of Sighs" for its constant marine accidents and subsequent shutdowns, had earned its nickname once again.

But the hero that day was bridge tender William MacDonald who, despite the risk to his life, sat calmly at his perch high above the bridge deck and quickly opened the span seconds before it would have been irreparably damaged. "The span swayed about six feet," he recalled. "We didn't know whether it was going to stand the strain. If we jumped one way we landed on the ship—the other way and we landed in the water. We sat tight."[7] Had MacDonald not acted so judiciously, the bridge might have faced a repeat of the September 19, 1930 incident when the log carrier *Pacific Gatherer* lodged itself under the ninety-metre fixed centre span, and as the tide rose, toppled it into the drink. The bridge was not rebuilt until 1934, leaving commuters to face either the ferry or, ignominiously, a barge across the inlet for four long years.

The *Bonanza* incident also highlighted the fact that the Lions Gate Bridge's three lanes were inadequate to handle current traffic demands, especially at peak commuting times. Given the approximate 400-percent increase in crossings in the ten years preceding the report, and the approximate 100-percent increase in the North Shore's population for the same period, the report estimated that traffic at both bridges was approaching "congestion proportions." Furthermore, despite access improvements and traffic control, the "practical capacity" of both bridges would soon be exceeded. Something had to be done, and quickly. It was a relief to many that the First Narrows Bridge Company had finally come to the table, but the crossing committee was recommending that not one, but two structures would be required by the mid-seventies: one at either end of the inlet. Both would need four lanes and combined would provide sufficient design capacity to carry traffic demands into 1976.

Small problem. There wasn't enough money to support the construction of two major structures if the government, under the Toll Highways and Bridges Authority, was to finance both. The Province had not yet ratified the First Narrows Bridge Company's offer. Despite the legislature establishing the Toll Authority as a Crown corporation in 1953, and authorizing it to borrow $10 million to purchase, construct and operate toll highways and bridges in the province, it still was not enough. The answer was to build either a new four-lane bridge at the First Narrows or a new six-lane bridge at the Second Narrows.

If *both* bridges were to be built, only four lanes would be required at the Second Narrows. But if only one was to be built, and the Second Narrows was to be the favoured of the two, then it would have to carry six lanes to replace the capacity of the old bridge as well as support growing traffic needs for the next twenty-two years. Tolls on either structure over the ensuing two decades would pay for one bridge at which point the other could proceed.

However, the First Narrows Bridge Company's plans for a third bridge were still on the table. By legislation, the company had the right to proceed with their bridge project, a position that appeared to be encouraged by Premier Bennett when he asked them "to get ready." But several weeks after the joint committee report was made public, the principals of the company issued a statement:

> Naturally, private investors must be provided with a comprehensive picture of the undertaking in which they are being asked to participate involving such matters as the present and future positions of the levels of government involved, the Public Utilities Commission and the Toll Bridges Authority . . . Until this has been done and time allowed for investors to give proper consideration, creation of controversy or pressure, no matter how well-intentioned, serves in itself as a delaying action, and does a disservice to the bridge-travelling public.[8]

The premier's refusal to allow a toll increase may have contributed to the skittishness of potential investors. The public was becoming nervous as well, after a local MLA warned that commuters would be paying tolls forever if another "private investor" structure was built across the inlet.

In hindsight, had the bridge company acted much earlier, they may have received the government's support. Minister Gaglardi had both the bridge company's offer in one hand and the crossing committee's report in the other. Although some committee members saw the two as complementary, and Gaglardi had classified a new First Narrows bridge as "absolutely necessary," he decided to refer the issue to his engineers for study—the death knell of another First Narrows bridge. The problem was how to land the structure in Stanley Park and where to carve out the route into Vancouver without further damaging Vancouver's jewel and adding more traffic to Georgia Street, one of Vancouver's busiest thoroughfares. Eyes began to shift east, toward the Second Narrows.

The report stated that during 1953, 5.6 million people and 3.2 million vehicles crossed the old Second Narrows Bridge, representing approximate 300 percent increases respectively for both people and cars over 1939 volumes. The increase for the same period over the Lions Gate Bridge was approximately four-fold for persons and fivefold for cars. But the Lions Gate did not have the frequent marine and rail interruptions that the old Second Narrows Bridge did. Without the frequent interruptions, would the Second Narrows route have been more popular to commuters? Likely. The 68 percent growth of traffic at Second Narrows over the two years prior to the study suggested that traffic patterns were beginning to shift as residential subdivisions and industry penetrated the eastern regions of North Vancouver.

Regardless of the location argument, at least there was now consensus that a new crossing was absolutely necessary. The report's conclusion was not only clear on that point, but alarming as well:

> It appears that all the Burrard Inlet Crossings will become saturated and inadequate to handle rush hour traffic volumes if the present rate of increase continues. The Lions Gate Bridge is expected to be congested by the end of 1955. The future of ferry service appears uncertain. The Second Narrows Bridge probably will be inadequate to cope with the volume of vehicular traffic within the next three years. With the general growth of traffic on the North Shore and the congestion of the Lions Gate Bridge after the fall of 1955, more traffic may be diverted to the Second Narrows Bridge. Thus the general growth of traffic would tend to result in a saturation of the Second Narrows Bridge at an earlier date. On this basis, it is anticipated the volumes during the rush hours on the Second Narrows Bridge might reach capacity within two years.[9]

The Committee's next step was to conduct surveys to determine what crossing capacity would be required for 1976, how the infrastructure should be divided between the two identified locations and then, when should it all be made available? Although the technical team set about to answer those questions, another more important one loomed: where was the construction money to come from?

Gaglardi's feuds with the federal government were well known. Although he proposed that both governments share BC highway costs equally—the federal government was already contributing to the province's portion of the Trans-Canada Highway in accordance with its two-lane standard—his requests were firmly denied. Meanwhile, to the general public and nagging opposition, provincial highway expenditures appeared to be out of control. In 1955, the year Bennett created the Department of Highways, its budget was $45.5 million, but a mere two years later it had swelled to $91.1 million. Gaglardi was on a mission to blacktop the province or, as he said, "take the wraps off British Columbia," and he had the blessing of the premier.

It was W.A.C. Bennett in fact, in his dual role as premier and minister of finance, who had control over the provincial purse strings, and he was anything but out of control. Bennett believed that economic prosperity would follow the roads. And despite the largess granted the Highways ministry, by the end of the decade Bennett still managed to eliminate the $190 million provincial debt he had inherited.

So, it was well within the government's agenda to finance a new Burrard Inlet crossing—in the interim, at least, until the travelling public could pay for it through tolls. It was also apparent that bridges were an important and necessary part of the province's highway construction plans. The province was not flat like the prairies and it was therefore necessary for new roads to traverse innumerable creeks and rivers as they climbed mountains and dipped into valleys and beyond. During the 1955–56 fiscal year, known as the "Year of Bridges," the Highways ministry constructed seventeen bridges and work was commenced on another fourteen. The next year they built twenty-four and work was started on another twenty-nine, and the year after that, 1957–58, it was a startling sixty-two with work commenced on another thirty. This level of construction was unprecedented in BC and perhaps even in the western world.

Being able to afford a new crossing, however, shed no light on where to put it. The technical committee's surveys explored various scenarios based on the estimated population growth of the North Shore which, in turn, was based on the land available for residential development. It all boiled down to peak-hour southbound traffic flows, which were estimated for 1976 at 4,040 vehicles per hour for the First Narrows and 2,000 for the Second Narrows. The 1953 peak loading capacity of both bridges was only 2,200 and 800 respectively, clearly inadequate to meet the future needs of the growing city of Vancouver and the municipalities on the North Shore. Although traffic pressure at the First Narrows in 1953 was almost triple that of the Second Narrows, it was expected to drop to roughly double that of the Second Narrows by 1976 as land became available for residential development at the eastern end of North Vancouver. In evaluating future peak-flow requirements, various options were considered.

A new four-lane First Narrows bridge, the very project proposed by the First Narrows Bridge Company, together with the extant bridge, was estimated to support either 5,200 or 5,900 cars per hour in the direction of main flow during peak periods depending upon how the lanes were managed. This would be adequate for 1976 and beyond. If a new facility was instead to be constructed at the Second Narrows (and it was assumed that the old Second Narrows Bridge would be inactive once that occurred), it would have to provide three lanes each way with a design capacity during peak hours of 1,300 cars per lane per hour. Together with the 2,200 at the existing Lions Gate Bridge, there would be adequate capacity for 1976 in this proposal as well.

Given the footprint and traffic-flow implications of building a new First Narrows structure, it was not surprising that the government announced the Second Narrows as the new crossing of choice. Neither Premier Bennett nor Public Works Minister Gaglardi had shown much enthusiasm for the First Narrows Bridge Company's offer, despite Bennett's comment to the company "to get ready." The general feeling was that the province and municipalities would be stuck with the bill to design and build the approaches while the bridge company would reap the benefit from increased traffic generated by the new highway to Squamish and the proposed superhighway to the border. And although the bridge company could not increase the toll rate, they could, under their agreement, continue to charge tolls until the end of their contracts. "Part of the Queen's highway system shouldn't be a private toll,"[10] Bennett argued. Although he was planning to charge tolls on the new Second Narrows Bridge, Bennett also believed that as soon as toll bridges are paid for they should be free to the public. It was likely this attitude, as well as Gaglardi's choice of bridge locations, that influenced the First Narrows Bridge Company's January 1955 offer to sell the Lions Gate Bridge to the Province.

From the municipal perspective, the Second Narrows was the best option, and the announcement was a relief to many, especially to Mayor Charles Cates of the City of North Vancouver: "As long as it is a good bridge, I think it will be satisfactory. For the City of North Vancouver, a Second Narrows crossing would be better than another bridge at First Narrows."[11] North Vancouver District Reeve Grant Currie echoed those sentiments: "It looks like our prayers have been answered ... It follows my argument all along that the provincial government should take responsibility for Burrard Inlet crossings. And also that a new Second Narrows crossing should precede a second Lions Gate Bridge."[12] Vancouver Mayor Fred Hume, who was just happy to have any bridge, neatly sidestepped the controversy by refusing to compare the merits of the two locations.

Others, like West Vancouver Reeve Jack Richardson, vehemently disagreed with the decision. After cheekily speculating that the provincial government might not be in power too much longer (they would be in charge for another seventeen years), Richardson suggested that the Second Narrows crossing might "produce major traffic problems in the east end of Vancouver."[13] He and his council could not accept the government's decision. In a two-page letter to Minister Gaglardi, he emphasized that the Second Narrows option was in opposition to the crossing committee's report, which stated that the Lions Gate Bridge would reach saturation before the Second Narrows Bridge, and that the majority of commuters were already using the First Narrows as a crossing of choice. G.E. "Bus" Ryan, president of the British Properties Home Owners Association, in an earlier letter to Richardson, had pointed out that more than 42,000 of the 60,000 people living on the North Shore lived west of Lonsdale Avenue in North Vancouver, and were more inclined to use the Lions Gate Bridge than the Second Narrows Bridge. But according to the City and District of North Vancouver Ratepayers Association executive, who favoured the Second Narrows option, North Vancouver residents used the Lions Gate Bridge out of necessity, not choice.

The government's decision was not made lightly, or without professional advice. In February 1955, Gaglardi had commissioned the experienced Vancouver engineering firm of Swan, Rhodes and Webster (later known as Swan, Wooster & Partners) to study the issue and recommend where to build the bridge. The leading partner of that firm, who would eventually be appointed by Gaglardi to design the bridge and consult on its construction, was Colonel W.G. Swan, a prominent Vancouver engineer. He was already well known to the government, having acted on behalf of the Province during construction of the Pattullo and Kelowna bridges.

Having cut his teeth in the CNR as a resident engineer, Swan soon rose to the position of district engineer at New Westminster where he was responsible for construction of the CNR mainland terminals on the Pacific Coast, including the Port Mann wharves and reclamation of the False Creek flats. When World War I was declared, Swan was already an officer in the militia regiment at New Westminster, and assisted with the formation of the 131st Battalion before being shipped to France where he served twenty-seven months as a major in the Second Battalion, Canadian Railway Troops, and as Light Railways and Tramways Engineer for the Second British Army. Thrice mentioned in dispatches, he returned home with a D.S.O., O.B.E. and the Croix de Guerre pinned to his chest—three awards that no doubt served him well throughout his civic career.

Between wars, Swan was busy building bridges. He was in charge of the design and construction of the Pattullo Bridge and the construction of the Lions Gate Bridge. It was on the Pattullo that he faced his greatest personal loss, one that he could not help but recall twenty-one years later when reviewing the roster of dead at the Second Narrows. His twenty-two-year-old son, William Mck. Swan, a prom-

Colonel W.G. Swan, a prominent Vancouver engineer with experience on both the Pattullo and Lions Gate bridges, was one of the most respected bridge engineers in western Canada. He was tasked with the responsibility of designing the new structure. PHOTO COURTESY UNIVERSITY OF BRITISH COLUMBIA ARCHIVES

ising University of British Columbia engineering student and gifted athlete, slipped on an oily girder while conducting a rivet inspection during a summer job. Plummeting thirty metres to the river flats, he died of internal injuries two hours after being rushed to the hospital at New Westminster. That Swan likely obtained the job for his only son must have been a burden that he carried with him for the rest of his life.

For men like Swan, losing themselves in their work was a form of salvation. It was probably a relief that two years later, at the outbreak of World War II, he was invited to Ottawa as director of construction of the War Supply Board. His tenure there was short lived, as he stated in a June 1940 press release: "The terms of my appointment with the war supply board in some respects were not acceptable to its successor, the new ministry of munitions and supply, and accordingly we agreed to part company."[14] Returning west, he settled at Work Point near Victoria, taking up the post of district engineering officer, and eighteen months later was promoted to Lieutenant Colonel. He would later be elevated to full Colonel, a rank that even post-war would still precede his name.

Colonel Swan's report, released by the provincial government on June 1, 1955, listed four reasons why the new crossing should be at the Second Narrows:

> 1) It would be cheaper than building a four-lane suspension bridge or tunnel, at First Narrows;

2) Rerouting would be needed for approaches through Stanley Park, in the case of a First Narrows location, and this would delay construction start, "a delay which the present (traffic) situation does not permit";

3) The present Second Narrows Bridge has a fifteen-ton load limit, which means heavier traffic channels through downtown Vancouver to the North Shore. If the new crossing were at First Narrows, it would mean that this heavier traffic would still go through downtown Vancouver, and thus not relieve the traffic load;

4) It will take only seven minutes longer to travel from downtown Vancouver via a Second Narrows new bridge, than it does now to travel via the Lions Gate Bridge.[15]

Not only was truck traffic through downtown Vancouver a huge consideration, but a bridge at the Second Narrows had an industrial benefit as well, which was one of Gaglardi's reasons for choosing it. The east end of North Vancouver was the only area on the North Shore available for industrial expansion, which a new bridge there would augment. Gaglardi, speaking on May 6, 1955, at the seventh annual tourist and highways conference sponsored by the Vancouver Board of Trade, in addition to emphasizing that point, cautioned that a new bridge at the First Narrows would create a bottleneck at Stanley Park. "If we need another bridge at First Narrows to handle traffic there in five years, then we can build another bridge at First Narrows. If that's not enough we can fill in Burrard Inlet," he voiced facetiously.[16]

In addition to recommending the Second Narrows as the preferred location, Colonel Swan's report outlined the costs of the various other options his firm had considered. His analysis revealed that the Second Narrows was not only the best logistical choice, but the cantilevered-style truss bridge was the cheapest plan at a projected cost of $11,594,040. Among the other options studied was a six-lane suspension bridge at the Second Narrows with a cost of $12,266,850, a four-lane suspension bridge at the First Narrows with a cost of $14,822,720, a causeway with locks and viaduct at the Second Narrows with a cost of $20,761,580, and a four-lane tunnel at the First Narrows with a whopping cost of $22,164,030.

The reference to a causeway at the Second Narrows was not a new idea, but in hindsight might have been the better alternative. It would have eliminated marine disasters at the Second Narrows as well as provided seagoing vessels with an opportunity to clean their hulls of marine growth in the freshwater lake created behind the causeway. First suggested in 1911 by the well-known civil engineer Walter Moberly, who at seventy-nine was still working, the concept drew on his many years of experience surveying the province, two of which were in the employ of the Canadian Pacific Railway seeking the best possible rail route to the west coast.

Moberly suggested that a causeway could not only carry the railway and a road for vehicular traffic, but would also permit access to ships through a system of locks. Two bulkheads, with wharves on either side, would support a lock for smaller vessels and one for larger vessels. Part of the plan involved moving the mouth of the adjacent Seymour River west by 800 metres to avoid siltation problems. Moberly believed that rail traffic to the North Shore would facilitate access to existing and untapped mineral deposits there, resources for a potential industrial base. His idea fell on deaf ears, but whether it was too farfetched or had more to do with unfortunate timing is uncertain, for on September 21, 1911, Laurier's Liberal government fell to Borden's Tories after fifteen years in power.

Major Swan, leading a small group of Vancouver businessmen, himself

LOCKS AND WHARVES AS ALTERNATIVE TO SECOND NARROWS BRIDGE SCHEME

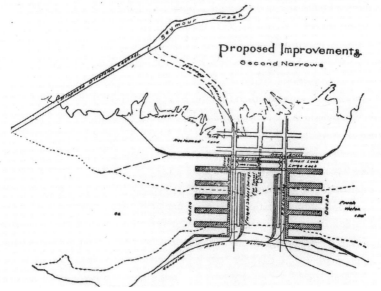

The causeway would have created a large freshwater lake behind it while still permitting tankers access to Port Moody. CITY OF VANCOUVER ARCHIVES

investigated the next attempt in 1928, even going so far as to draw up detailed plans including costing out the proposal at $8.5 million. His recommendation was for a six-hundred-metre causeway carrying an eighteen-metre roadbed flanked by a double-tracked railway line. Road traffic, which would go over the ends of the locks, would be unimpeded by ships passing through the locks. After due consideration by all levels of government, the plan was duly scrapped.

But good ideas don't die easily, and twenty-three years later Port Moody revived the plan by requesting that the National Harbours Board investigate it. The Guinness family, who were a few years away from selling the Lions Gate Bridge to the province, was reportedly studying it as well. Two months later, Vancouver City Council jumped on board and on February 25, 1952, at a Burnaby-Richmond Federal Liberal Association gathering, Liberal MP Tom Goode promised to bring the matter up with the federal department of transport on his next visit to Ottawa. Dr. W.D. Bains of the National Research Council, who authored the 1954 federal study, reported that it would be less expensive and more efficient than a bridge at the same location. First Narrows' tides would be reduced from 6.3 knots to 1.8 knots and access to Port Moody would no longer be fraught with danger. That the idea died again was perhaps not unexpected given the complex nature of the proposal and the numbers of voices needed to consent to it, including several federal departments as well as the provincial government. But it is not the last we would hear of the causeway option, though the next study would place the crossing farther to the east.

Colonel Swan, who was now in favour of the bridge option, given that he had studied all the alternatives, stated that the main objection to the causeway was its cost, now double the estimate of the new bridge: "I don't believe such a plan would be cheaper than the bridge ... For one thing, a 108-foot viaduct would have to be built to get traffic over the locks and the ships passing through them. This would probably cost $10,000,000 in addition to the $15,000,000 for the causeway, locks and necessary dredging. Without it we would have the same problem as we have now, of traffic being held up every time a ship goes through the narrows."[17]

Although the Vancouver Merchants Exchange were concerned that building the bridge would defer potential causeway plans by years, Colonel Swan added that if the government still wanted to proceed with the causeway option at a later date, strategic positioning of the new bridge piers would permit that. As well, the height of the bridge would also not impede the loftiest of tankers from accessing Port Moody.

Now that the Second Narrows had been officially revealed as the crossing of choice, a curious public wanted to know exactly where the bridge would sit. Gaglardi, however, was keeping that close to his chest. He had reasons for delaying the announcement: at first he didn't want to inflate property values in the area, and later balked at releasing details because property purchases for the approaches and cloverleafs had not yet been finalized. In hindsight, it was obvious where the bridge was not going to be. To the east of the old bridge was the BC Electric Company's Bridge River transmission line, to the east of that were six water mains delivering water to the Greater Vancouver Water Board, and to the west of the old bridge were the berthing facilities of the Alberta Wheat Pool. The location was therefore constrained by previous development that would be difficult to disturb.

More than likely, Gaglardi just didn't want to alert the public to the acrimonious expropriation process that had residents calling the government "high handed" and "fascist." Gaglardi was in the middle of twenty-four separate negotiations for land purchases that at the end of the day cost the province only $178,224—an average of just over $7,400 each, and a small fraction of the total bridge cost. Co-operative Commonwealth Federation MP Harold Winch and CCF MLA Arthur Turner, both elected in Vancouver East, criticized the provincial government's handling of the land expropriation process. They wrote letters to Premier Bennett and Minister Gaglardi complaining that Highways personnel had no credentials, surveys were being conducted without the land owners' permission and some residents still had yet to receive their notices of expropriation. Two weeks later, Gaglardi swore that he had yet to receive the letters, but advised that if homeowners were unhappy, an arbitration process was in place to address any grievances. One property owner, incensed at the approximate $14,000 the government offered for his home, hired four independent appraisal firms, the lowest of which was $4,000 over the government offer and the highest $12,000 better.

As early as May, speculation that the new bridge would be located about forty-six metres west of the old bridge was confirmed by drill tests completed on the south shore to determine whether the bottom there was suitable for sinking piers. Drilling was expected to start during the week of May 16 on the North Shore. By the middle of August, years of bitter public debate were drawing to a close when seven test piles found adequate footings for the new piers. Construction could now begin.

3

Substructure

An engineer is a very staid, solid individual who deals with
facts but without imagination . . . an engineer is not required to
have vision. Why does an engineer need vision? All an engineer
needs to know is what is the strength of a piece of steel . . .
　　　　　　　　　　　　　　　　　　　　　　　–PHIL GAGLARDI
(TOURIST AND HIGHWAYS CONFERENCE, MAY 6, 1955)

Although Gaglardi initially expected to begin the tender process for Contract
No. 1 in the early summer of 1955, it was not until October 8 that tenders for the
substructure—the concrete piers to support the bridge—were finally announced.
Completed bids were required no later than noon on Tuesday, November 29. It
was expected that the project would take about two years to complete. Simultaneously, bids were being let for the remaining 6.9 kilometres of highway between
West Vancouver and Horseshoe Bay, and plans to connect the bridge to the new
highway were also underway. When the print media pointed out that tenders
had already been announced, but that the exact bridge location had still not been
revealed, Gaglardi instructed his chief engineer Neil McCallum to release the
details.

　　　The new bridge was to be launched from the south shore just west of Koote
nay Street with approaches leading from Skeena, Cassiar and Cambridge. It
would be located just west of the old bridge on the south shore and angle across
the inlet to land 120 metres west of the old bridge on the north shore. There, it
would cross Main Street via a low-level overpass and cloverleaf east and west, accessing the Dollarton Highway and Main Street respectively. A new road would

The old Second Narrows Bridge's days are numbered as the south-side steel launches just west of it on the south shore. COURTESY JIM ENGLISH

direct northbound traffic to Keith Road and Lillooet pending completion of the connecting highway.

The bridge, which would rise 62.5 metres above tidewater, with about 44.2 metres of shipping clearance, would be a six-lane "deck type" cantilevered-truss structure with the main span stretching 335 metres of the total 963-metre length, two anchor spans and four deck truss spans making up the difference. Altogether, including approaches, the bridge would be about 3.6 kilometres long, and would have the distinction of being the longest cantilevered-truss structure in western Canada, the second longest in Canada, and the eighth longest in the world. The "deck type" refers to a bridge with the roadway supported by steel while older cantilevered spans generally have the roadway running through supporting steel. It was essentially a continuation of the highway, with the same look and feel of one, and was comparable in most aspects to the Aurora Avenue Bridge, an eight-hundred-metre structure carrying traffic in and out of downtown Seattle.

There are bound to be cost overruns on any major project, but the approximate $2.5 million increase before bridge construction even got underway was staggering. When the plans were submitted to the federal Department of Transport, they were rejected on the basis that the distance between the centre-span piers was insufficient for safe navigation, even though it was just under three hundred metres. The centre span would have to be lengthened by about fifty metres, thus adding to the projected cost. Swan, the designer, was perplexed at the decision. Gaglardi appeared to conveniently forget the catastrophes at the Narrows when he complained in the legislature that "The old Second Narrows bridge is only three hundred feet. Why do we have to have this extra length? I never heard of a ship that was that wide. This has increased the cost of the bridge by several million."[18]

Cost overruns would get worse, but attention was momentarily diverted to the bidding process. By the end of November 1955, the substructure bids were in and it was announced that the joint bid of Peter Kiewit Sons Co. and Raymond International Corporation Ltd. was the lowest at $4,314,369.70. It was almost a million dollars less than the highest bid, but well over the $2 million estimated by the province to complete the job.

The two companies, both American with Canadian subsidiaries, were anxious to begin work, but before their joint venture could be confirmed as the successful bidder, McCallum had to review each submission. A week later he announced Kiewit-Raymond International Ltd. as the successful bidder. Raymond International would manage the project. The company had until August 31, 1957, to construct the piers before the ironworkers would begin to erect the steel. By the end of January 1956, enough land had already been cleared for offices and storage. Raymond President J.A. Dickenson advised that all the required machinery would be purchased locally and that all the employees, except for five American engineers and supervisors, would be Canadian. One of the Americans, project production manager Judson Howell, arrived in Vancouver on January 23 from Mobile, Alabama, to begin the process of hiring men and purchasing equipment for the estimated mid-February start.

Simultaneously, in a government move to appease disgruntled Lions Gate commuters still smarting over Gaglardi's choice of bridge locations, bids were being accepted for a new east-west access cloverleaf and a bridge over the Capilano River to ease congestion there. In a somewhat opportunistic bit of political manoeuvring on the eve of a Vancouver Centre by-election, Premier Bennett even hinted at a new First Narrows bridge if the planned six-lane Second Narrows structure failed to address the North Shore's traffic woes. Gaglardi, in fact, reported that plans for the new structure would be prepared within the next three years, and by the middle of February 1956 had already reserved 11.3 hectares just east of the Capilano River for the proposed north bridge landing. He cautioned, though, that the government had no concrete plans to proceed at that time. The land would be utilized as a gravel pit until it was needed.

While Kiewit-Raymond International Ltd. was busy constructing a trestle that would jut out into the inlet to facilitate the pier construction process, a drama was being played out in Victoria with Gaglardi and McCallum. The previous November, Gaglardi had announced that he was about to reorganize the Highways department with its ten dysfunctional engineering districts—districts that operated independently of one another and that occasionally tripped over one another in the discharge of their duties. Gaglardi had already made a radical change by setting up the BC Toll Highways and Bridges Authority, whose "mandate was to construct, operate and finance crossings in the province with the Department of Highways conducting surveys, choosing sites and supervising construction under the new Minister."[19] The premier himself chaired this working body, known simply as the Toll Authority, whose membership included Gaglardi, other ministers of the Crown and a few other prominent government officials.

The reorganization, which would create four distinct geographical regions—Kamloops, Nelson, Prince George and Vancouver/Victoria—had been inherited from the previous government but was deferred until the Socreds settled in. The planned restructure was now urgent, but there were other changes that Gaglardi wanted even more. As the political head of the ministry, he didn't like the fact that his chief engineer, who was also chair of the Highways Board, was making important decisions. McCallum was equally irked that Gaglardi was making decisions that undermined his authority. The conflict first arose when the board disagreed with one of Gaglardi's decisions. Gaglardi then wanted to know why he wasn't chair of the board. After being informed by McCallum that as chief engineer he was the "ex officio" chair, Gaglardi adjourned the meeting so that they could look up the meaning of the phrase. Once he discovered that it meant "by reason of the position," Gaglardi took action to amend the statutes, thus relieving McCallum of his decision-making duties, but leaving his position as chief engineer and salary intact.

That the minister had a narrow view of what engineers did was very apparent, an opinion he had no difficulty expressing publicly. He had only achieved grade seven himself and it was evident that he was still gripping the levers of a virtual D8 Cat. He seemed to reserve his greatest respect for his ex-cronies: contractors and the men who ripped up the earth to build his roads. He commented to the press that engineers merely follow formulas in books and are therefore wholly without vision. His outspoken nature, a healthy part of his controversial persona, often got him into trouble. His son Bob sheds some light on why he thought like that: "He would just devour everything and anything, because he thought he was inadequate, because he didn't have education ... you know, his thirst for knowledge was never-ending, so he had opinions on everything. He had his own views and he didn't mind to tell you."

On February 1, 1956, two department heads—Doug Willis, chief paving engineer, and Joe Cunliffe, assistant paving engineer—quit the ministry to start their own paving company: Willis, Cunliffe and Tait. By itself this was not unusual, but two days later Neil McCallum stunned the government by offering his resignation as well. Fingers pointed to Gaglardi's decision to build a tunnel at Deas Island to cross the Fraser River. After a three-year traffic study, McCallum's planning committee had recommended a four-lane crossing at Annacis Island, a four-lane crossing at Port Mann and an additional four-lane crossing at Annacis. McCallum viewed Gaglardi's decision as a vote of non-confidence and believed that he had no alternative but to resign.

When pressed in the House about the departures by Opposition leader Arnold Webster, Gaglardi tried to deflect the focus by suggesting that they were due to low government wages and that the province could not possibly match salaries paid by industry. A week later Lloyd Willis, divisional engineer at Penticton, departed after being informed that he would be demoted to district engineer because of the reorganization. Willis's comments, after announcing his departure, shed new light on Gaglardi's salary argument: "The conditions under which we're expected to work have become untenable."[20]

On February 24, the House erupted in anger at the rapid departures of so many senior highway personnel—as many as fifteen in the previous few weeks—and wondered what Gaglardi was doing to cause such "discontent and friction."[21] For four rowdy hours, the Opposition railed at him, finally launching a ruddy-faced Premier Bennett to his feet to defend his scrappy Minister. Members of the Opposition demanded that Gaglardi table the resignation letters. Arthur Turner, CCF MLA for Vancouver East, claimed that "dark clouds are hanging over this government," but the thick-skinned Gaglardi merely brushed aside the criticism as "hogwash."[22]

A couple of weeks later, Gaglardi did table the letters, revealing McCallum's discontent with the reorganization and his failure to be consulted as the reasons for his departure. McCallum's comments went further, however: "We have lost several valuable members of our organization who feel it is hopeless to try and carry on, and therefore I have no alternative but to resign from the position of chief engineer which, under existing conditions, has become untenable for me."[23] By the end of March, the hole in the Highways Department had been patched with sixteen new engineers. Eight resumes were also on the minister's desk for the chief engineer's job, but the shift change essentially gutted the department at a critical stage in the province's infrastructure development—and right at the start of the construction of the new Second Narrows Bridge.

Meanwhile, in Vancouver, a disgruntled group of East Hastings residents was facing down the government over what they considered to be unfair land purchase tactics. Although most were in favour of the new bridge, en masse they were complaining that government offers for their homes were well below private appraisals and that the government was employing high-pressure tactics to force them to sign. Their letters, phone calls, and even visits to Victoria were being ignored. Some complained of being intimidated by survey crews warning them that they had better settle soon as the deadline was fast approaching.

On the north side of the inlet, Tommy Johnson, a Native who went by the name of "Chief" Tommy, was having his own little war with the government. Seventeen years earlier he had relocated from the Mission Reserve a few kilometres to the west to an 8,000-square-metre site right in the path of the bridge. He had been offered only $10,500 for his house, outbuildings and orchard despite a government appraisal of $18,000. Gaglardi, who appeared to be reviewing every detail of the land purchases personally, refused to believe his own appraiser and sent another to revalue the property. Complicating matters, Tommy's land was part of a larger eight-hectare parcel described as part of No. 2 Indian Reserve, that the government was in the process of expropriating for the north bridge landing. Claiming that he had been granted full right to the land, Tommy was adamant that if the Province failed to pay him what he thought he was due then they would have to smoke him out. Although the Province eventually paid the Receiver General, who represented the interests of First Nations people, $95,000 for the eight hectares, Chief Tommy's personal share of that remains unknown.

Also in the path of progress was the nearby Canadian Swedish Rest Home, pastoral in every sense when it was first built in 1948 on Cutter Island at the mouth of the Seymour River. Once shaded by a nearby forest, it would soon find itself isolated in the middle of a constricting cloverleaf after a small arm of the Seymour River was filled in, erasing Cutter Island and creating an environment as hostile to respite as one could imagine. The Province expropriated the property, offered its owners $251,000 in seemingly uncontested compensation, and promptly moved the highway district into the empty building. It was soon apparent, however, that not all the residents had departed. Employees working late into the night would nervously recount stories of creaking floorboards and the telltale scuffing of slippers across cold floors.

What little solitude remained in the area soon departed when Russ Hubensky, a driver for Deeks-McBride Ltd., drove his Reo model F-506M mixer truck onto the construction site in February 1956 with the first load of concrete, thereby kicking off almost two years of frenetic activity. Evans, Coleman and Evans, a BC concrete company, would supply approximately 35,000 of the nearly 60,000 cubic metres of concrete required for the whole bridge, which it would mix from

Each pier, 2 through 10, hosted an east and west column that rose up to support a massive concrete pier cap crossbeam upon which sat the bridge deck. OTTO LANDAUER OF LEONARD FRANK PHOTOS, JEWISH MUSEUM ARCHIVES LF-35079

as many as 460,000 sacks of Elk Portland Cement supplied by the BC Cement Company.

By the middle of June, Kiewit-Raymond, who had up to 150 men working feverishly on the project per shift, had already built five of the required fifteen piers on the north side of the inlet and had applied to the federal government under provisions of the Navigable Waters Act to start construction of those located below the high-tide mark.

There would be seventeen piers altogether: fifteen on the north shore and two on the south shore. Five of the north shore piers would be below the high-tide mark, as would one on the south shore. The first ten on the north shore would therefore be built on solid ground underlain by boulders, heavy gravel, sand and silt, much of it naturally conglomerated (essentially glacial deposits and outwash materials from the slopes high above the inlet). Pier 1, which comprised a concrete abutment wall, was built over steel piles. But given the natural compaction of the soil on the north side, piles were not required for Piers 2 through 10. Instead, spread footings were poured up to eight feet beneath the surface to support the pedestals and columns and help to distribute the weight of the bridge and traffic to the soil beneath.

Each pier, 2 through 10, would host an east and west column that would rise up to support a massive concrete pier cap crossbeam upon which would sit the bridge deck. Columns would also rise from Piers 11 to 14, but the steel would sit directly over these columns without the need for concrete crossbeams. The columns of Pier 14 were much larger, and hollow, to support the anchor assembly for the north anchor span (a similar mechanism would be installed in Pier 17 as an anchor to the south anchor span). Anchor spans were required to carry the tension generated by the loading of the long, clear centre span. No columns were required for Piers 15 and 16, upon which would rest the steeply curved ends of the centre span.

Each of the columns was reinforced vertically with No. 11 rebar—a type of heavy reinforcing steel—at fifteen-centimetre centres and horizontally with No. 6 rebar at thirty-centimetre centres, with the steel descending right down through the pedestal and deep into the footing. On top of the huge pier cap crossbeams—Piers 2 through 10—rested the concrete stringers that would form the base of the approach spans' roadbed. At the top of the columns that would carry the steel on the north side, wells and chambers for cantilever anchor-arm tie-down assemblies were built. A tie-down assembly is a crucial structure used to support a cantilevered span while it is being built. It will be discussed in greater detail in chapter 5.

To celebrate the construction progress, on July 12 Social Credit leaders turned up en masse to a concrete-pouring ceremony. Health Minister Eric Martin playfully threw his old hat into the wet concrete, declaring that "part of me will ever be with this bridge."[24] At the same gathering, Gaglardi announced that the projected $16 million cost of the new bridge was looking a little light given that a steel strike in the United States had forced up the price of steel. Premier Bennett, also in attendance, announced that he was wiping out tolls for *passengers and pedestrians* on the new Second Narrows Bridge, Lions Gate Bridge, Oak Street Bridge, Agassiz-Rosedale Bridge, and those at Nelson and Kelowna, thus saving the travelling public about $400,000 a year. He called his tax relief a "Second Dividend" for the travelling public, the first one being the government's acquisition of the Lions Gate Bridge, thus preventing decades of private tolls there.

Contradictory to the progress shared at the concrete-pouring ceremony, work on the bridge was temporarily stalled when it was discovered that the Province would have to seek permission from the Burrard Inlet Tunnel and Bridge Company to construct Piers 5 and 6. These piers would be located on the part of their land used as an approach to the old bridge. As well, four years earlier, the bridge company had contracted a twenty-year lease of the land to the CNR until 1973, when the CNR would take it over free of encumbrances. As the leased land now fell under the auspices of the Railway Act, their authority had essentially been extinguished. Although Reeve Grant Currie and Mayor Charles Cates had both expressed a huge sigh of relief when the new bridge had been announced, now they were ironically right in the middle of the delay.

According to Colonel Swan, however, it was the company's lawyer, Dugald Donaghy, who was the holdup despite the CNR's offer to grant the construction company temporary access pending completion of the paperwork. Although the two mayors claimed there was no hostility, the delay would hopefully play into any compensatory claims they might make "for water mains, lighting and the north approach road as well as for loss of revenue and inconvenience to the bridge company."[25]

Mayor Cates announced that the bridge company was going to operate the old bridge with a lower toll rate than the new bridge, which Cates denied was a reprisal against the government. Under the agreement with the CNR, the bridge company continued to collect the tolls. "I don't care what they charge," Gaglardi said. "They can run it free for all I care. All I hope is that it doesn't fall down before our new bridge is completed."[26] Ironically, it would not be the old bridge that would suffer that indignity.

It was not until June that the Province realized it could not just expropriate land that fell under the Railway Act, and had to apply for permission to the Board of Transport Commissioners in Ottawa to build on bridge company land. Kiewit-Raymond had already built Piers 1 through 4 and went ahead working on 7 through 9, but the fact they would have to haul their equipment back to complete Piers 5 and 6 was going to create delays and add to cost. The application was further delayed when it was summarily returned to Victoria due to a bit of bureaucratic silliness that only two governments can generate. Victoria had applied on behalf of the BC Highways Department rather than the BC Toll Highways and Bridges Authority, and it took another month to affect the change. When it was finally authorized, it came with one condition—that the BC Toll Authority build a short stretch of road around the proposed structures to permit access to the old bridge; a small price to pay for the convenience of a new crossing.

While it had been waiting for federal permission, the joint venture company continued to construct the rest of its north-side piers, the last five of which were to be built below the high-tide mark. Building bridge piers in the middle of watercourses has challenged bridge builders since the Romans first invented the process in the second and first centuries BC. The question of how to build a stable structure in the middle of a river when the current impedes work and the bottom is usually unstable was solved by these innovative bridge builders out of necessity. It was also fitting, perhaps, that the plucky Minister of Highways was of similar ancestry (*sans* the engineering credentials), and had been referred to as such by the premier. Piers 11 through 15 would therefore have their footings in the water and would require dewatering prior to construction. Pier 10 also required dewatering due its close proximity to the inlet, but it would be high and dry once built.

There are a couple of common methods engineers use to dewater a site for the construction of a bridge pier. The first is the *caisson*, from Old French meaning

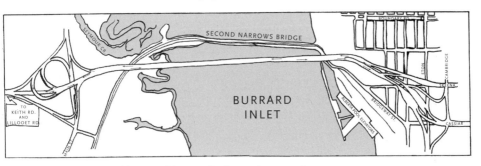

Proposed layout of the new bridge and approaches. ILLUSTRATION BY JOAN JAMIESON, RECREATED FROM A DIAGRAM THAT RAN IN THE *VANCOUVER SUN* ON OCTOBER 14, 1955

"large box." Although there are several types of caissons, the one mentioned here is the pneumatic caisson, only because it was made famous by the builders of the long-serving Brooklyn Bridge in New York City. A pneumatic caisson is simply a box that is open at the bottom to facilitate work on the river or seabed, and closed at the top so that the space within the enclosure, once pumped dry, can be pressurized to keep out mud and water, and to permit men to work within the enclosure. There were two caissons built for the Brooklyn Bridge, one on the Brooklyn side and one across the East River on the New York side. They both had working shafts—water shafts for transferring excavated material to the surface and supply shafts. A much less sophisticated version of this device was used to construct Pier 16 of the Second Narrows Bridge.

The second device, older and perhaps less complex, that engineers employ to dewater bridge pier sites is the cofferdam, from the French word, *coffre,* again meaning "box." Cofferdams were invented by the Romans to facilitate the building of stone piers in watercourses. The first cofferdams were constructed of log piles driven in a circle to depth by blows from a rudimentary pile driver. Additional log piles were then driven inside the first pile circle and the gap between the two ranks packed with water-impervious clay. Water within the enclosure

A modern cofferdam, consisting of high-strength metal sheet piles, was used to construct those bridge piers built below the waterline. OTTO LANDAUER OF LEONARD FRANK PHOTOS, JEWISH MUSEUM ARCHIVES LF-36344

was evacuated by bucket before the riverbed was excavated to the required depth. Timber piles were then driven into the exposed riverbed and cut flush to form the base of the stone pier.

It was this method, though modernized, that was employed by Kiewit-Raymond when constructing Piers 10 through 15, the latter pier of which faced the full force of a very powerful tide. Instead of timber piles, however, high-strength metal sheet piles were employed to form the cofferdam. Manufactured in a variety of styles, sheet piles are essentially about sixty centimetres wide by fifteen to eighteen metres long and heavily corrugated for strength. A male and female coupling at either edge interlocks with its neighbour, thus providing support as well as sealing the completed structure against the intrusion of water, mud and debris. Even then, in a highly pressured environment, water will squirt in from innumerable infinitesimal cracks.

Typically, the process of installing a cofferdam first entails dredging the seabed to remove soft surface mud. Further dredging is undertaken to dig a "glory hole" before strategic piles are driven to support a steel frame, within which sheet piles are driven in a large rectangle to form the cofferdam. In the case of Pier 14, this rectangle was approximately 19 metres wide by 42 metres long. A pit was then excavated to the required depth, approximately 1.5 to 2 metres beneath the mud line, within the cofferdam. In the approximate 800-square-metre area described by the enclosure, 808 capped timber piles—capped to prevent damage while driving—were driven in a 90-centimetre-by-90-centimetre spacing arrangement to *practical refusal*, a term referring to the point at which there is a risk of damaging the pile if driving continues (essentially, when the pile meets dense material).

The piles were then cut off 90 centimetres above the bottom of the excavation and on top of the pile nest, thirty-five concrete trucks, accessing the site from the adjacent wooden trestle, poured a seal course of dense tremie concrete (heavy concrete delivered under water) through two 30.5-centimetre tremie pipes to a depth of five metres. A tremie pipe is a device through which the heavy concrete is delivered, and is sometimes referred to as an "elephant's trunk." The 3,058-cubic-metre pour, which was the largest bridge pour in western Canada at the time, involved four plants producing concrete for twenty hours straight. The concrete was then left to cure for three or four days before the cofferdam was dewatered. Once dry, the project was treated like any normal construction site where forms were built to complete the balance of the pier including the pedestal and columns. Divers later cut off the sheet piles at the tremie level.

Piers 11 through 14 were all constructed in this manner using timber pile foundations prior to pouring the tremie concrete. Pier 14 was of particular importance in that it was the north anchor pier. Pier 15, known as the north main pier, was also critical given that it would support the long, clear centre span, the span that makes this bridge an important structure. Special measures were therefore required to construct it, not only because of what it would support, but also because of its location and the tidal forces that would be exerted upon it. Resting

The massive columns of the north anchor pier, Pier 14, under construction. OTTO LANDAUER OF LEONARD FRANK PHOTOS, JEWISH MUSEUM ARCHIVES LF-36517-A

in 7.8 metres of water, the pier, which measured approximately 20 metres by 38 metres by 13 metres deep, was built within a cofferdam, shored and braced to withstand the 151.5-tonne longitudinal force experienced during a typical five-knot tide. Inside the cofferdam, 538 steel pipe piles, 31.5 metres long and weighing 50.5 tonnes each, were slammed to depth by a McKiernan-Terry steam hammer before the piles were filled with sand and capped with concrete. Over the pile nest was then poured a 3-metre slurry of tremie concrete. The pier itself was twenty-one-celled with 60-centimetre-thick interior cell walls and 90-centimetre-thick outside cell walls. Cross drains were installed to accommodate fluctuating tidal elevations.

Pier 16, the south main pier, was similarly important in that it would support the south side of the long, clear centre span. A type of caisson was employed to build this pier, which would sit directly on the hard sandstone substrate beneath the south side of the inlet. A 3-metre-thick caisson was built of tremie concrete over a false bottom—which made it buoyant—within the cofferdam of Pier 14 before it was finished. Once the pier was dewatered, the tremie concrete was levelled with additional concrete before a side of the cofferdam was removed to permit the caisson to be floated out at high slack tide. It was then docked next to the trestle and built up before the 5,443-tonne behemoth was towed to the Alberta Wheat Pool wharf on the south side by five tugs pending the right tide to submerge it into a blasted sandstone crater 30 metres north of the south shore. Once resting on its bed of 10-centimetre-diameter crushed rock 30 centimetres deep, a tremie seal ring was poured around the base of the structure to stabilize it and render the caisson watertight. The false bottom was then removed and an island formed around the site from excavated material to facilitate access. The caisson was then filled with 6.6 metres of concrete, further stabilizing it before the pier was finished much like Pier 15.

Pier 17, the south anchor pier, was perhaps the most complex of the eleven piers built over land. Burrowing down through over 15 metres of overlying glacial conglomerate, excavators carved out a 4.5 x 21 x 15.3-metre shaft right down to sandstone. Slumping due to heavy rains necessitated moving the excavation back into the hillside with the aid of box shoring to prevent further deterioration, and due to the awkward location, concrete had to be delivered over a specially constructed road with a crane and bucket. Inside the pier was built a similar anchor assembly to that installed in Pier 14.

Although building this pier was a challenge, it was not half as difficult as Minister Gaglardi's simultaneous duel with Ottawa over the federal government's potential contribution to the bridge construction cost. If successful, he would

be able to offer the public toll-free travel on the new bridge and then by necessity, tolls would be lifted on the Lions Gate as well. Gaglardi was hoping Ottawa
would classify the bridge as a "difficult" gap under the Trans-Canada Agreement,
which would make it eligible for 90 percent federal funding. This was also the
position of the federal Minister of Fisheries, James Sinclair, who lived in West
Vancouver.

Under the normal federal formula, Ottawa would pay 50 percent of the cost
of two bridge lanes, which would see the Province stuck with the remaining 50
percent of the federally funded lanes plus the full cost of four lanes. Under this
plan, the province would be lucky to be funded for four to five million of the
total bridge cost, and as the bridge would then become part of the Trans-Canada Highway system, it would be exempt from tolls. In the end, it was just not
worth it and the federal offer was abandoned in favour of the public paying the
freight. Minister Sinclair accused the premier of withdrawing the bridge from
the Trans-Canada formula just so that he could preserve the $1.25 million in toll
revenue from the Lions Gate Bridge, but the premier denied the accusation.

While the remaining piers—14, 15 and 16—were in various stages of completion, Kiewit-Raymond was also busy producing the 108 pre-stressed concrete
beams that would hang between Piers 2 to 10 to form the roadbed of the approach spans. They were made of pre-stressed concrete: concrete that is prepared
with heavy wire cable under tension. Tension was applied by jacking the cable in
the opposite direction to the load it was to carry and then locking the tension in.
Pre-stressed concrete was necessary because it has ten times more capacity to carry tension than regular concrete. Bridge members undergo tension or compression, or both, depending upon where they are located in a bridge. Pre-stressing
concrete can replace steel in some instances, is often cheaper, and can be made
onsite to any desired shape.

Each beam, of which there would be twelve between each pier, was 36.6
metres long by 2.1 metres high, and weighed 87.7 tonnes. The longest concrete
beams produced in North America at the time, they were manufactured onsite in
metal forms at a rate of one a day for 108 days. Six vibrators, four on the outside
and two on the inside, compacted the concrete, and within each beam were set
ten 4.06-centimetre-diameter Tecon Kopex tubes through which ran the wire
for pre-stressing the concrete. The wire was then jacked to the required load and
grout forced into the tubes to lock in the stress. Once the beams were cured,
a crane lifted them into place where four diaphragms were poured along their
length to link them together.

Despite construction progress at the Second Narrows, traffic demands at the

It was calmness in the face of danger that allowed the men of Local 97 to hang the steel of the new Second Narrows Bridge. PHOTO BOB DOLPHIN

First Narrows continued unabated. On Sunday, August 4, 1957, there was a 2.5-hour traffic snarl and two four-car pileups on the Lions Gate Bridge, prompting a frustrated Vancouver Mayor Hume to ask the premier and the Highways minister what they planned to do about it. "Even if they started tomorrow on plans for a new crossing it would take five or six years for completion," he fumed. "The Lions Gate Bridge is out of date now and by then it will be completely overloaded."[27] Premier Bennett's non-committal reply that the government was studying the problem and had been for years was in direct contradiction to Gaglardi's recent statement that no studies were underway or planned.

Carl Stanwick had other thoughts on his mind as he strolled the wooden trestle adjacent to the piers on a mild Saturday in early July 1957. The resident engineer with Swan, Wooster & Partners, he had graduated from UBC as a civil engineer in 1951 after having spent almost two decades in the construction industry, and was now supervising the Second Narrows Bridge construction. Content that everything was on schedule, he commented, "Things are going well. We should have this contract finished in September. Then the steel structure contract people can go to work."[28]

The bridge was already swarming with the ironworkers of Local 97. This new breed of men was entirely different from the substructure contractors. For them, risk was not only an obligation, but a signature of their profession. It was calmness in the face of danger that set these men apart, that made them appear simultaneously careless yet mindful. It was, in fact, their ability to stride confidently across whisper-thin strands of steel high above Burrard Inlet that allowed these men to hang the steel of the new Second Narrows Bridge.

4

The Bridge Company

*Dominion Bridge was born of the same impulse as modern
Canada itself, inextricably wrapped up in the dream of Sir
John A. Macdonald, Canada's first prime minister, to unite the
nation with a transcontinental railway...*
　　　　　　　　　–DOUG FETHERLING, *VISION IN STEEL*

Just to the south of the northwest bridge on-ramp, and approximately half a kilometre west of the Ironworkers Memorial Second Narrows Crossing, sits the Lynnwood Inn, known as the "Lynnie" to many of its loyal patrons. With part of the
structure dating to 1935, it is one of the oldest working hotels in British Columbia.
Muddy pickups and tradesman's vans surround the building on almost any given
day. The meals are tasty and substantial, and the number of quilted shirts and dirt-
fingered ball caps slouching in the age-worn booths attest to its popularity. On this
particular January 2008 morning, the frigid rain slashes sideways before a cutting
wind from the west, gusting over a group of grizzled smokers huddling around the
entrance. The glowing tips of their smokes wink protectively in calloused hands,
and despite the weather, their raspy laughs and friendly banter greet this day like
any other. It is, after all, just another workday. Right behind the building, the rumbling cadence of an early-morning freight lumbers through the air.

　　Lining the walls of the lobby and the cafeteria are the sepia-toned photos of long-departed patrons, a phalanx of dockworkers, shipbuilders and
construction gangs, and in one corner, a pictorial shrine to both the old and
new Second Narrows bridges. Staring from one small portrait is the visage of
a youthful worker. Holding a pair of caulk boots and wearing the fluorescent
hardhat and trademark grey Stanfield's wool undershirt of a faller, he seems

The men who built the Second Narrows Bridge. COURTESY LOU LESSARD

curiously out of place amongst the bridge photos around him. No one here knows his name or anything about him; he and the men in the other grainy photos are now mostly gone, names and identities lost to the daily routines of other men like himself.

After work, allegiances easily switch to the pub next door, one of the last of the true beer parlours: wooden tables, scarred with age not fashion, sturdy wooden chairs, the pervasive whiff of beer, and friendly barkeeps calling out to regulars. Although this bar was built only in 1967, it is in the similar, original pub, now the Lynnwood's beer and wine shop next door, where the men who built the Second Narrows Bridge gathered to review their day and too often to pay respects to their dead. Ironworking has always been a dangerous profession and sometimes getting lost in a glass or two is a necessary prescription. They will always insist that it is gravity that is the real killer, not the steel.

These were the men of the International Association of Bridge, Structural, Ornamental and Reinforcing Ironworkers Union Local 97, a chapter of the International Association of Bridge and Structural Ironworkers of America. Ironworkers in BC were initially managed through Local 86 in Seattle, Washington, but it eventually lost the franchise due to distance and logistical factors. W. Atkinson, together with a group of fourteen other men, applied for and was granted the Local 97 charter on May 11, 1906.

The early part of the twentieth century was a strategic and often violent time for North American unions, who were trying to find their rightful place in an oft-unjust society. Union struggles in BC, although less violent, were little different. On May 2, 1910, twenty-seven unions formed the BC Federation of Labour out of dissatisfaction with the national Trades and Labour Congress. Although the switch did little to allay the damage caused to wages and job security by the concurrent depression, it did provide unions with a unified voice.

It wasn't until World War II that a surfeit of opportunity was presented to the metalworking trades, which gratefully plied their craft building the many Victory ships required to transport men and equipment to Europe for the war effort. Membership in Local 97 grew and activity even improved following the war. Despite the collapse of the shipbuilding industry, the province hovered on the cusp of a major construction boom that would last well into the fifties. It was within this economic bubble that a new bridge across Burrard Inlet was envisioned and built. In anticipation of this, on October 1, 1956, the BC Toll Highways and Bridges Authority announced that tenders were being let for Contract No. 2, the steelwork. Bids would be valid if accepted no later than noon on November 7, and as quickly as a day after the process closed, Dominion Bridge Company Ltd. (DB) was announced as the low bidder: their price, a hefty $11,099,672. Together with the substructure contract, increasing the length of the centre span, painting and lighting, the bridge cost had now mysteriously swelled to a surprising $19 million—approximately $7,400,000 over Colonel Swan's original estimate and as much as $2.5 million over Gaglardi's recently published figure. It would get worse.

That the low bidder was DB was a given. The giant steel-erection company, with its roots in the heart of Quebec, had built many of Vancouver's early landmarks, and one would be hard pressed to drive anywhere in the downtown area without seeing a DB project. The elegant Hotel Vancouver and the Marine Building, both built in 1929, were two; the Pattullo Bridge, built in 1937, another. The Lions Gate Bridge followed a year later. DB built the General Post Office in 1955 and then the BC Electric building in 1956. A year after the bridge collapse, the company would finish the Lions Gate Hospital on the North Shore. The Port Mann Bridge would follow in 1963 and the unique West Coast Transmission Building in 1969. The name "Dominion" was fitting. The company literally began a few years after the birth of the nation, first as the Toronto Bridge Company operating out of Toronto, and then, in 1882, as the Dominion Bridge Company with a second location on a 9.5-hectare site at Lachine, Quebec.

The company's prospectus, issued on September 23, 1882, formalized its purpose to construct bridges for the Canadian Pacific Railway. As the CPR united this vast nation with a ribbon of steel, so too did it nurture this pivotal company with a legion of bridges. In 1904, the company began its operation in BC with a 732-metre combination railway/road structure across the Fraser River at New Westminster. Known as the Fraser River Swing Bridge, it was a double-decked affair with vehicles crossing on an upper deck and rail traffic beneath. Although the Pattullo Bridge replaced the road function of the bridge in 1937, the original rail structure, minus its upper deck, is still being operated and maintained by the CNR today.

Lions stand guard at the entrance to the Lions Gate Bridge, constructed by Dominion Bridge for the Guinness family in 1937–38. OTTO LANDAUER OF LEONARD FRANK PHOTOS, JEWISH MUSEUM ARCHIVES LF-22503

By 1930, the company was firmly established in the province, operating from an approximate twenty-four-hectare yard at 2150 Boundary Road in Burnaby. This swampy patch of land, northwest of Burnaby Lake and beside the tracks of the Great Northern Railway, would become known as "The Plant." Due to the instability of the boggy soil, yellow cedar piles had to be driven to support the heavy steel fabrication machinery. Although the equipment is now long gone, the large machine pads still stand proud against the fissured asphalt. It was here that parts of San Francisco's Golden Gate Bridge were fabricated as well as the Lions Gate Bridge, the Second Narrows Bridge, and many other structures in between.

Steel was delivered to the site in 3.7 by 6.2–12.3-metre sheets of plate steel, wide-flange beams, eye-beams and angles by rail to the main fabrication building, a colossal 1.6-hectare steel and wood-roofed structure that at times bustled with hundreds of men. The steel, which was referred to as "mill ordered material" because it related to a specific job, in contrast to "stock material" ordered for general use, was manufactured to specification in England before making its perilous way by ship across the Atlantic, through the Panama Canal and north to Vancouver. In the northeast sector of the building, a 27.2-tonne crane facilitated the fabrication of smaller members, while in the main chamber of the structure, a massive gantry crane mounted 15 metres above the concrete floor moved heavier pieces along the fabrication process. In this industrial cathedral, the enormous chords—a term referring to the top and bottom members of a truss—were cut, punched, riveted and welded. Some weighed as much as 96.5 tonnes. DB had also subcontracted some of the fabrication work to Western Bridge, which was based in a smaller but similar plant nearby at 145 West First Avenue.

The Boundary Road facility was not DB's only plant. At the time the bridge was being fabricated, the company had nine thousand employees making anywhere from $1.68 to $2.41 per hour working from fourteen such facilities across the country. After celebrating its seventy-fifth anniversary in 1957, it was embarking on a $20-million expansion that would increase its capacity by 40 percent within the next two years. Allan S. Gentiles, vice-president and manager of the Pacific Division since its inception, stated that the company already had "more work on the books for 1958 than its total 1957 production."[29] Although DB had begun life as a bridge builder, by 1958 bridges represented only 10 percent of its total productivity nationwide. Despite that, it was still known locally as "the bridge company," even though there was another Vancouver firm with the same informal name—the company that owned the old Second Narrows Bridge.

As the fortunes of steel erection companies began to wane, DB sold its Boundary Road property in the mid-seventies to explore opportunities south of the border. A portion—six hectares—of the parcel was picked up by BC Hydro and BC Transit, but it was the incidental use of the site as a film studio that would eventually see the preservation of the company's original buildings, specifically a portion of the fabrication plant, the cafeteria, the compressor building and the boiler plant. Although the fabrication building was scaled down from its original size to a modest 4,000 square metres, the steel skeleton of the remainder of the original structure still stands as a testament to earlier times. Now, the brightly painted red steel, exposed to the elements, speaks more of art than industry.

In 1987, the Province assumed control of the site under the provincial Crown corporation PavCo to preserve the film studio business, which was becoming an important part of the provincial economy. It has operated as Bridge Studios ever since.[30] At one end of the main fabrication plant, in stark contrast to its earlier use, sits the facade of a medieval village, home to the *Stargate Atlantis* television series. But walk the length of this building, known as the finishing shed, and it isn't films that come readily to mind. Close your eyes and you can still hear the clang of steel, the hiss of compressors and the shouts of men. Through the miasma glows the cherry-red heads of rivets, welders' torches arc brightly in the gloom and fountains of sparks spray brilliantly from a myriad of grinders. The acrid stink of industry stings the nostrils. In this environment of metal, heat and sweat, the bridge was born.

Not far from the plant was the bridge itself, where Kiewit-Raymond had finished the piers and the four approach spans, setting the stage for DB to begin its contract. By early November 1957, fifteen ironworkers were on the site, preparing to erect the steel. Steelwork had hardly begun, however, when the ironworkers experienced their first fatality. On December 6, Albert Bearchell, a twenty-eight-year-old ironworker and member of the bolting-up crew with three to four years of experience, had just finished lunch with his younger brother, Roland. After attending a weekly safety meeting, he was carrying an eighteen-kilogram scaffolding bracket across a nineteen-centimetre ribbon of wet steel just behind his brother. As they drew abreast of the traveller, a crane used to lift the steel, Albert complained that his load was digging into his shoulder and that he was going to shift it. The bracket caught on the housing of the derrick, knocking him off balance.

His startled yell attracted the attention of big Jim English, the job superintendent, who watched helplessly as Albert fell and struck the cross bracing before hitting the bottom chord approximately fifteen metres below. There he hung precariously before plunging another fifteen metres to the ground. Roland turned away at the last second, not wanting to witness the awful truth. He would quit

Fabricating a massive gusset assembly at the "Plant" on Boundary Road, now the site of a major Vancouver film studio. DOMINION BRIDGE CO. LTD.

ironworking soon after. English rushed down to comfort the mortally wounded Albert, who died a few hours later, leaving a wife and three children. Little did English know that he was witnessing a smaller version of a greater tragedy to come.

This incident renewed the age-old discussion that had probably been argued on an infinite number of bridges and high steel structures—that of the use of safety lines and nets. Walter Miller, a Workman's Compensation Board inspector, when asked at Bearchell's inquest about the use of nets, stated that it was not standard practice. This had also been the conclusion of an October 29, 1956, meeting between the WCB and a number of senior DB personnel, whereby it was determined that nets would be of little value to the erection and bolting-up crews because it would be impossible to string them until the steel was hung. Four days later, in a letter to DB erection manager John Prescott, WCB corroborated the consensus that nets would be impractical.

Otmer C. Carpenter, the general construction superintendent on the bridge, had had a hand in almost every major bridge construction project in BC since

Left to right: High Carpenter, general construction superintendent; Jim English, job superintendent; and Don Jamieson, field engineer. PHOTO BOB DOLPHIN

joining DB in 1929. Known to all and sundry as "High" because of his lofty stature and his love for heights, Carpenter was dead set against the use of nets as well. When working on a bridge in the US that was dozens of metres above a gorge, safety nets had been installed as a precautionary measure. One day High asked the foreman how the nets were working. "Well, come on, I'll show you," he said, as he jumped off the bridge and into the net. High fired him on the spot, citing that the nets brought a false sense of security to the job. "That's not why they're there," he argued.[31]

Carpenter, an American from Spencer, West Virginia, was well qualified in the subject of safety. In 1913 at the age of twenty-five, when his shoulders made contact with two bare high-voltage wires at the top of a Eugene, Oregon bridge, 11,500 volts ripped through his rugged frame. He plummeted twenty-six metres, cannoning off the steel in an uncontrolled freefall to the Williamette River. After three years in hospital, a badly reconstructed nose, no voice box and a withered hand, when a friend in New York suggested that he leave the heights and join him in the trenches building sewers, High's whispered response was probably typical of most ironworkers: "I would sooner fall off a bridge and be killed instantly than die a lingering death at the end of a muck stick."[32] After being similarly maimed, most would find another occupation, but not High: he went right back on the steel that eventually made him one of the most respected high steel bosses in North America. "I've seen a lot of accidents on jobs," he recalled forty-five years later, "and I have never seen one yet when looking back, that couldn't have been avoided. Safety is common sense and doing things the right way. It's as simple as that."[33]

Some of the men called him a "prince," no faint praise for an ironworker, especially one in charge, but his bosses called him a "gentleman." That High was a bit of a legend amongst DB executives was very apparent. Once, when John Prescott asked him for the receipts to back up an expense report, High replied: "I count my money when I leave for a job, and I count it when I get back. The difference is what you owe me."

The bridge was proving to be a dangerous place to work. Three deaths had already preceded Albert's. First, there had been Alex Hauga, who was electrocuted. Then, on February 22, 1957, it was Alexander Robertson, a member of a crane crew employed by Kiewit-Raymond, who was struck on the head by a heavy section of timber and died overnight in North Vancouver General Hospital. A couple of months after that it was Syd Belliveau, a driller who drowned when his tug overturned in a heavy riptide. Now it was the ironworkers' turn with Albert. In comparison, during the building of the Lions Gate Bridge twenty years earlier, only one fatality was experienced when the earth in a construction pit gave way, crushing a worker.

Jim "Skip" Pratt, DB's construction administrator, was domiciled in one of the mobile buildings on the north bridge approach. He remembered that there were many non-fatal accidents as well, such as broken arms and legs, that were costing the company time and money as well as elevating its WCB rating. The company was fully compliant, and more, with WCB regulations: a qualified first-aid attendant was on site; a fully equipped first-aid room had been set up on the north bridge approach; weekly safety meetings were held despite WCB only requiring monthly meetings; all the men wore hard hats and life vests; a 7-metre rescue boat with two 35 hp Evinrude outboard motors patrolled the inlet; and a safety man, experienced ironworker Verne Sidelman, had been appointed despite WCB not requiring it. But accidents were still occurring far too frequently. The young Australian assistant field engineer, John McKibbin, when writing home to his sister on April 22, 1957, about bridge building conditions in Canada, commented that "They think nothing of losing ten lives on a job. I was horrified to see the conditions under which the workers work."[34]

It was a serious problem that needed addressing, and in an effort to stem the injuries and resultant losses, the company sent Pratt and Roy McWaters to San Francisco to learn from the giant steel erection company, Bethlehem Steel Corporation. DB imitated their program, including site inspections, more safety meetings, photos and instructions to report unsafe practices before they became an issue. It was only after these new procedures were implemented that DB's accident record and WCB rating began to improve. However stringent the safety precautions were, though, unknown factors still lay in the individual actions of the men and the inherent risks of the job itself. Both were incalculable.

Don Heron, twenty-two years old in 1958, remembers what it was like to dance the iron. Weaned on the stories of his father and grandfather, both ironworkers, he had little chance of another occupation given that ironworking was almost a genetic disposition of his family. It was this way with many ironworking families, where the temptation—perhaps obligation is a better word—for sons to follow their fathers was irresistible. Once, in fact, all three generations of Herons worked on the same job together. When Don was punking in the fifties, even the slightest hesitation or intimation of fear would attract the derision of his fellow ironworkers, most of it in fun. Early in his career, on one job a superintendent asked Don to do him a favour.

"Donny," he called out, "I want you to go down there and put the centre

Don Heron, ironworker, knew what it was like to dance the iron. COURTESY DON HERON

bolts in the star bracing. You sit on the ball and take a bag full of bolts and go down there and bolt them up."

Don hesitated a second before asking, "Do you have a choker or something I can tie off with?"

The super laughed and said, "What's wrong, you scared? You chicken or something? What's wrong with you?"

Don wrapped his leg around the ball, and was lowered gently to complete the task, one of the most dangerous jobs in the erection process. When he was finished, the super signalled the operating engineer running the crane to jiggle the ball a little on the way up. It was a mischievous little white-knuckled initiation rite that Don now chuckles about, but that would never happen today. After that, it was standard practice for him and the other connectors to ride the ball down off the bridge to lunch and back up again with a load of steel, punctuating the air with, "Let's do some tonnage." Like monkeys, they scampered unencumbered all over the steel, scooting up and down the columns, scuffing their boots as they found tenuous grips on flanges and rivet heads. "You'd wear out your boots fast, you know—you'd rip the stitching," Don recalled. Walking along the bottom flange of a beam while gripping the top with your hands was also a popular method of getting around.

"Now you got two guys in a basket to connect the star bracing, one at the top and one at the bottom," he says. "When I was a connector, I did that myself. The bracing had come up, I make the bottom end and, you know, shinny up the column to make the top end and slide down and say, 'Well, let's move on.'"

Now, ironworkers follow the stringent regulations of WorkSafeBC (formerly WCB), which has devised a whole lexicon around the field of working-from-height safety. Read any WorkSafeBC guideline for workers employed at heights, and it is peppered with terms like "fall arrest system," "fall restraint system," "fall protection system" and "full body harness." In essence, ironworkers will say— some with slight contempt—that it is a 100 percent tie-off requirement, meaning that they cannot move anywhere unhindered on the steel. Being tethered has removed some of the romance and bragging rights of the trade. Bridge workers were once revered for their feats of daring; now the job has become less risky and more mundane.

The loss of that freedom, however, has meant a sharp decline in fall-from-height fatalities in the few short decades since the bridge was built. Although there were ten such deaths—excluding the bridge collapse—in the fifties, thirteen in the sixties and fifteen in the seventies, they declined to seven in the eighties, six in the nineties and remarkably, no such deaths were reported between 1998 and 2006. This does not mean that ironworking has become a risk-free occupation. The profession still attracts a one-in-six chance of injury, greater than that of both miners and fallers.

Perhaps some of the old ways began to change with the construction of the Second Narrows Bridge. The bridge was unique in that it was one of the first major structures in western Canada to be bolted rather than field riveted. Field riveting was a dying art by the time Don Heron entered the scene in 1957, and it would soon take its last breath. Don remembers working with his father John on a riveting gang. The first day on the job was a revelation for the young punk.

"You call me Johnny, don't call me Dad, and you're Don and, you know, don't do any son, dad stuff. I'm not going to do you any favours. I'm going to heat up one extra. If you miss one and it goes in the river, I'll give you one miss and then you're down the road."

"And I said, 'Well, you're not the boss.' He says, 'Yeah, but you're my son and I'm not going to have you foul up like that!'"

A riveting gang was usually composed of four men—a heater, a catcher, a bucker-up and a driver—who could process up to four hundred rivets a day. The heater, who heated the rivets on a small coal-burning forge, more or less ran the gang. He would ask how many rivets were needed and what sizes, and then he would pluck the red-hot plugs from the forge with tongs, knock off the scale and fling them like a fastball to the catcher who would snag them in a metal cone.

"You know, my dad should have been a baseball player," Don recounted. "I guess I'm bragging a little bit, but all the heaters got that way. They got a certain way of holding the tongs and firing—it's a totally lost art now. And they could fire it up to, you know, thirty feet, forty feet, right into the cone . . . Bang!"

The searing chunk of soft rivet steel would then be excised from the cone by the catcher with his tongs and inserted into an available hole right up to the rivet's buttonhead. The bucker-up would then place a tool, known familiarly as a "horse cock," over the head of the hot rivet and hold it tightly while the driver flattened the other end with a pneumatic hammer. Inside a chord, instead of a horse cock, an Air Jam was used, or if space was tight, the bucker-up would choose a Ben Joe. All of this activity was carried out on a platform of narrow planks suspended from the steel with rope. One wrong move and it was *adios amigos*.

Don almost experienced that terror first-hand when he was working 60 metres above the water with a partner, Eddy. They had just finished the job and were taking up the floats, "and usually what you did is that you'd unhook the two outside lines and then you'd pull the float up on the inside. Well, one line kind of jammed and I went down like a damned fool to try to kick it with my foot and the whole float flipped down so I was hanging, holding onto the edge of the top of the angle of the float, and Eddy reached down—because he was a big boy in those days—and he reached down and grabbed my arm and pulled me back up onto the beam and he said, 'Don't do that again,' and I said, 'I promise. I won't.'"

Although the major elements or members of the bridge were still riveted at the plant during the fabrication process—985,000 rivets were used—the individual pieces were assembled in the field using high-strength *quenched* (cooled in water or oil during manufacture) and tempered steel bolts. Four hundred thousand were required, thus making the bridge the largest bolted structure in Canada at the time. The technical term for the style of bolt was ASTM A 325, the acronym referring to the American Society for Testing Materials. Bolts were delivered to the north end of the bridge by truck where an old ironworker, Stan Marshal, sorted them by size into different buckets that he stored in a shed until they were needed by the bolting-up gang at the front end.

The advent of bolts meant that a new gang had to be established. Where once there were heaters, catchers, bucker-ups and drivers, now there was a gang that did nothing but bolt the raising gang's work. Using impact wrenches fed from a topside compressor, the bolts were tightened from the outside while another man crawled inside the large chords, plumb posts and diagonals through entry ports to hold the nut firm with a spud wrench. The deafening chatter of the impact wrench, amplified by the steel, dulled their hearing and rattled their teeth in the confined spaces, which required them to spell each other off at regular intervals.

Each bolt that was tightened had been planned to the last detail. In fact, any bridge is first envisioned in an engineer's office, and later minutely detailed in a drawing office by experienced draughtsmen. Called "shop detail drawings," they illustrate the fine points of each member so that a shop foreman knows where to punch holes, weld and paint. Essentially, these drawings are the engineer's

Making the connection. As a chord is manoeuvred into position by the traveller, the raising gang, clinging to the steel without safety lines, wrestle it into place.
PHOTO JIM PRATT

instructions to the fabricators on how to build a particular piece of the bridge. Although DB had a drawing office at the plant, it was standard practice on a major project to establish an independent office to handle the details. One of the reasons for this was that DB was very busy and had a tendency to steal a man here and there to assist with other projects. The establishment of an independent detailing shop made that practice more difficult and kept the detailers that were assigned to the bridge focused on the bridge. Unknowingly, this practice of isolating major projects would prove to have fatal consequences.

Hugh Dobbie, a twenty-year-old immigrant from Scotland, was one of the detailing crew at the temporary office set up on Hastings Street. Coincidentally, he lived nearby on Dominion Street and had applied to both Dominion Directories and Dominion Bridge when he first arrived in Canada in 1956. He got calls from both companies, each offering $165 per month. He chose Dominion Bridge because it was only a fifteen-minute walk from his home, and began, with seventeen other young recruits, to learn the painstaking work of detailing steel-work: "You had your project leaders and your squad bosses as you called them then, and you got your detailers and you had your checkers and your checkers checked all the work. It generally worked out that for every three or four detailers you had one guy checking all the work and he would send you back until it was 100 percent correct and once it was correct it would go down for approval to the engineer of record."

The engineer of record was Swan, Wooster & Partners, the consulting engineer, though on the bridge Murray McDonald was the engineer in charge. He was a UBC honours grad who had been with DB for twenty-one years, the last twelve in the erection department. According to vice-president and manager of the Eastern Division, Robert Eadie, McDonald was the company's second-most senior erection engineer in all of Canada and was "noted particularly for his thoroughness and safety mindedness."[35] Assigned to the bridge as field engineer by John Prescott, McDonald was excited by the prospect of running his own project. Prescott remembered inviting McDonald into his office.

"Do you want to be the engineer?" he asked McDonald.

"Yes," McDonald quickly replied.

"Well, get the drawings and start looking it over. You're going to have to build some falsework."

McDonald's son Paul and daughter Ann recall that "he used to come home at night, and he and Mom would sit down and have a beer and talk before din-

ner, every night, and I guess that's what they talked about, how his day had gone. Because he was, at the end there, I remember, very thrilled that he was going to be in charge of that bridge."

On November 25, 1957, McDonald prepared a memo entitled, "Proposed Organization and Duties," at the direction of his boss John Prescott. A note at the bottom stated, "The above defines duties in general, but no

Murray McDonald, known as "High Pockets" because of his lofty stature, eagerly accepted the challenge of building the new Second Narrows Bridge. COURTESY PAUL MCDONALD

engineer would be restricted to these duties." McDonald, known as Engineer "A" in the memo, was to be responsible for: "Administration, expediting, sub-contracts, rigging and equipment, cost control, liaison with consulting engineers and inspectors, safety."[36] McDonald's real duties, however, included a much broader interpretation of that note. He was not only to be the field engineer, but would wear two other weighty hats as well, that of the project's erection engineer and design engineer—major project responsibilities that today would be assigned to others. Heavily burdened, he was badly in need of an assistant. He called John Prescott, who was in Montreal on business at the time, for permission to hire a young Australian, John McKibbin, whom he had recently interviewed and liked. John Prescott recalled the conversation.

"I've got an application from a young Australian, John McKibbin," he said. "I need an assistant. May I hire him?". McDonald asked.

"Yes, go ahead," Prescott replied. "I'll meet him when I get back."

The young engineer's May 28, 1957, letter home about landing the job with DB was telling: "All the trouble I had to get a job, one would imagine I would end up with the most dead end job of all time. But not at all, this company I'm working for has so much work in its lap that it could do with another dozen engineers, yet one has got to just about plead with them for a job in their lousy outfit. I'm working on the erection of a ten million dollar bridge which will be going up in August. So with three months to go before erection starts, they have given me the job of thinking how it is going to be put up. I shudder to think at what is going to happen if this casual attitude is the general thing. At home there would be a team of one hundred engineers of top calibre for ten years before. Anyhow no one is worried so why should I? Money is no object here."[37]

Known as Engineer "B" in the memo, McKibbin would be responsible for: "Surveying, including layout of work, and checking all chainages and elevations. Supervision and records of piledriving, construction of falsework footings. Grouting and setting anchor bolts. Preparation of drawings for falsework and temporary structures. Job diary."[38]

McKibbin had graduated from Sydney University in Australia two years earlier, and had one year of general experience while attending school and a couple with the Cleveland Bridge Company prior to arriving at Vancouver with his wife, Barbara, on what Australians like to call a "working holiday." He and his new bride were immediately struck by the beauty of the local mountains and pursued skiing with a passion. It was his plan, all along, to leave Canada for England in another few months before returning to his homeland. This plan might have been upset, however, by DB's intention to assign McDonald and McKibbin to another challenging structure, the high-level Alexandra Bridge across the Fraser Canyon, immediately following the completion of the Second Narrows Bridge. It was a challenge that McKibbin would probably have had a hard time resisting.

McKibbin had landed at the right firm to further his engineering experience. DB was "large and very engineer oriented and almost all of the department heads

any other way the way things are here.
Well love to all , one and all

John

THIS IS SECOND NARROWS
BRIDGE. 20,000 Tons of Steel
1100 ' Main span.

John McKibbin's excitement about building the bridge was evident in the sketch he drew for his family after landing a job at Dominion Bridge. SKETCH BY JOHN MCKIBBIN, COURTESY PETER MCRUVIE

were engineers with the exception of the comptroller who was a CA ... the sales manager was an engineer, the construction manager was an engineer, the general manager was an engineer, the chief engineer, of course, the shop, plant superintendent was an engineer; everybody who was in a highly responsible job was an engineer."[39] But there was something less attractive about the company as well, something almost sinister. Nepotism was ripe in the large steel erection company, and whether you were an engineer or not, at times it was so incestuous that you had to be careful whom you talked to and what you said for fear your comments might come back to haunt you. Progress through the ranks was often a matter of whom you were related to or whom you knew, which made it difficult for many engineers waiting their turn for leadership.

One of those in the wings was Cliff Paul, the designer of the large travellers and the third engineer assigned to the bridge. Paul, who was domiciled in DB's engineering department at the plant, worked directly for chief engineer Angus McLachlan. Known as Engineer "C" in the memo, he was responsible for: "All problems concerning detail and design of permanent structure. High tensile bolting, stud-welding, painting. Gas Lines and trams. Shop and drawing errors. Engineering supervision of erection procedure. Quality control."[40]

A quick glance at the two field engineers, McDonald and McKibbin, one middle-aged and one young, and one could be forgiven for thinking that all engineers are cast from the same mould: tall and slim. They could even have been

mistaken for father and son. McDonald, known as "High Pockets" because of his towering stature, was 6' 4" and 165 pounds, while McKibbin, equally lean, was 6'. Where McDonald was taciturn and serious, McKibbin was loquacious and charismatic, but they balanced each other nicely and the fit became a solid working relationship. McDonald, at fifty-one, already had a lifetime of bridge experience by the time the two met, which to the

Murray McDonald, one of Dominion Bridge's most experienced erection engineers, in his office. COURTESY PAUL MCDONALD

John McKibbin, assistant erection engineer, upon graduation from Sydney University, Australia. COURTESY PETER MCRUVIE

twenty-three-year-old McKibbin was a brilliant opportunity to learn.

Soaking up knowledge like a sponge, he wrote of McDonald in his last letter home on May 21, 1958: "My boss' name is Murray McDonald—a real whiz on bridges—he has had a hand in nearly every bridge in the Northwest. So no publicity please. He seems to delegate me a lot of responsibility on the job, far more than I ever had at Liverpool and I like it, but it needs an old hand at the helm to know all the local intricacies, and there are lots of things common here that I've never heard of."[41] He would soon absorb enough information from McDonald to qualify to register as a Canadian engineer.

Canadian engineers wear an iron ring on their pinkie finger, a sign of their profession. The ring is part of a decades-old tradition that would appear to any outsider to be more of a secret society than anything to do with the solemn occupation of engineering. Reciting ritualistic oaths at closed meetings called "Camps" overseen by "The Corporation of Seven Wardens," the Ring Ceremony would sound Tolkienish if it were not so noble. Wearing the iron ring on the right pinkie finger, if one is right-handed, is an engineer's right and privilege, one designed by the man who brought us *The Jungle Book*, Rudyard Kipling. Kipling was tasked with the responsibility by the Engineering Institute of Canada of designing a ceremony to remind young engineers of their responsibilities, and it was he who envisioned the iron ring as a symbol of that code of ethics. "It should be rough, as the mind of the young man. It should not be smoothed at the edges, any more than the character of the young. It is hand-hammered all around—as the young have their hammerings coming to them," he wrote.

Myth and legend endure longer than truth, and one myth that needs dispelling is that the first iron rings were cut from the steel involved in the 1907 and 1916 Quebec Bridge disasters, where a total of seventy-three men died. Although a compelling story, it is nothing more than that, as is the myth that Camp No. 5

(Vancouver) rings are cut from the twisted steel of the Second Narrows Bridge. Both these bridges were constructed of steel, not the raw iron from which the rings are made.

It was something more tangible than myth, however, that occupied McDonald and McKibbin as they pushed the steel south. Work was progressing smoothly, and a well-oiled system was in place to keep the bridge advancing toward its conclusion. There were deadlines to keep.

5

Ironwork

What more epitomizes the Ironworker than his ability to pull it all together in the face of great adversity and get the job done. I think about the jobs we have all done going back one hundred years. Most jobs are a challenge in one way or another, whether an engineering feat, or enduring the extremes of the elements, or fearless acts or mind-numbing heavy work. Look about you. The infrastructure of our country was built by the ironworker and we are always at the sharp end of every project. Yet we always look forward to the next adventure and reminisce fondly about the ones in the past.

—PERLEY HOLMES, *CONQUERING THE CHALLENGE*

I imagine that bridge engineers are an exact bunch that has little time for supposition. Their belief system is intricately tied to their ability to calculate the stresses in and on a particular structure; otherwise, it may not stand up. Although there are many examples of structures that have not survived the slide rule or computer, compared to the sheer volume of structures that do, the failure rate is very low. Of course, whenever there are failures, especially ones ending in human fatality, any rate of failure is considered too high, and quoting statistics to prove otherwise is little comfort to the bereaved.[42]

Although I am not an engineer, and therefore cannot provide a rigorous analysis of the forces bearing on a steel structure, a rudimentary description of them is necessary to understand how the steel of the Second Narrows Bridge was hung. The real and potential forces influencing any bridge are many and varied. Gravity, the weight of the structure itself, the traffic loads it will bear, wind and

Ironworkers scramble over the steel still hanging in the slings of the traveller. They will quickly make the connection from the wooden plank floats hanging from either side of the chord they are perched on. DOMINION BRIDGE CO. LTD.

snow loads, are some of the forces that engineers consider that we would all understand. Then there are forces that engineers call dynamic loads, which refer to sudden, one-off cataclysmic events such as tsunamis, tornadoes or earthquakes. The stresses placed upon a structure during one of these events are difficult to predict and therefore to design for. Witness the overpasses that collapsed during the 1989 San Francisco earthquake.

Two key forces that are not so apparent to the casual observer, but are inherent in every structure, are tension and compression. In fact, bridge members are always under one or the other or both. Tension is when an object is being pulled apart or stretched, while compression is when it is being pushed together or shortened. Some materials are better at being compressed than tensed and vice versa. Concrete is a material that is better at compression than tension and that is why it is often pre-stressed to improve the latter.

When a load is placed upon a beam, or bridge, it bends or bows downward. However imperceptibly, the top part of the beam is forced together and experiences compression, while the bottom part of the beam draws apart and undergoes tension. Too much compression and the beam buckles, too much tension and it snaps. It is the responsibility of bridge designers to anticipate these forces and dissipate or transfer them so that no one part of a bridge bears a disproportionate amount of stress.

Read a few engineering books on bridges and your head will spin with the number of types and designs they discuss. For our purposes, however, a mention of four basic types will suffice: beam, arch, suspension and cable-stayed. A beam bridge is perhaps the simplest, although it can be made more complex— and stronger—with a multitude of truss options. The problem with simple beam bridges is that they can only span short distances compared to the other three examples. The arched bridge, which is a good example of a structure that can dissipate force, can link much longer gaps; while a suspension bridge, like the Lions Gate Bridge—which is an excellent example of a structure that can transfer force—can bridge even longer gaps. The road surface of a suspension bridge is hung from catenary cables supported by towers and anchored at either end. The loads that bear down on the bridge deck from gravity and traffic are transferred via these cables to the cable anchors and towers and from there into the ground.

Cable-stayed bridges rely on a multitude of cables fanning out from a tower to support the roadbed. Although cable-stayed bridges may look like suspension bridges in that the roadbeds of each structure are supported by cables, the difference is how the cables are connected to their towers. In a suspension bridge the cables ride freely across the tops of the towers while in a cable-stayed bridge they are anchored to the towers, thus transferring loads directly through the towers and into the ground.

With respect to the arch, it is one of the most ingenious tools in the bridge engineer's pocket. Although the Sumerians invented the corbelled arch (a crude form of stepped arch) in the 7th century BC, it was not until the 2nd century BC that the Romans smoothed out the design for bridge and aqueduct use. Their early arches, which were semicircular in design, were built with wedge-shaped bricks or stones. A keystone at the apex of the arch locked the structure together. The ingenuity of the arch is its ability to dissipate the forces acting upon it by directing them outward and downward along the curve of the structure and into the ground via the supports at either end. An arch is always under compression.

There are several Roman stone and brick arch bridges, or remnants thereof, still surviving today. But perhaps the most amazing historic arch, and one more fitting to our story, is the Ironbridge arch spanning the River Severn at Shropshire, England. Built of cast iron between 1779–1781 by Abraham Darby III, it was his grandfather, Abraham Darby I, who in 1709 perfected a method of smelting iron with coke, thereby cutting production costs and kick-starting the industrial revolution. That the bridge still stands and is used today is a testament to the ingenuity of the early engineers who designed it and the early ironworkers who erected it. Perhaps the reason the bridge still exists is that cast iron has wonderful compressive abilities, which the form of the arch accommodates, but is brittle and therefore weak at tension. Push it together and it holds; pull it apart and it snaps.

Many would be surprised to learn that the Second Narrows Bridge, despite its arch-like shape, is not an arched bridge in the truest sense. In fact, it is a fancy variation of a beam bridge known as a cantilevered truss with drop-in, the latter term referring to its suspended centre span (more about this in chapter 12). Despite the bridge not being a true arch, parts of it act like an arch with respect to the forces acting upon it. Where the variable depth truss (deeper truss over the piers) meets the piers, the lower chords are under compression as is the nature of a true arch, but in the middle of the span the forces are reversed, with the lower chords experiencing tension and the upper chords experiencing compression as is the nature of a simple beam.

The Second Narrows was obviously too wide for a simple beam bridge, but was ideally suited to either a cantilevered truss bridge or a suspension bridge (cable-stayed bridges were not common at the time) both of which were considered by the design engineers, Swan, Wooster & Partners. That they chose a cantilevered truss was largely because it was the cheaper of the two options, but it would also provide motorists with an unobstructed view of the harbour to the west, Indian Arm to the east and the North Shore mountains to the north. Cantilevering is simply described as a beam or girder that is anchored at one end and free at the other. It permits engineers to design longer spans with variable depth trusses as they did for the Second Narrows Bridge. In that case, two cantilevered arms were launched from the north and south shores, to be connected in the middle of the inlet by the drop-in span.

Despite the longevity of the Ironbridge at Shropshire, it was not cast iron that would be used to build the Second Narrows Bridge, but a mixture of structural carbon steel and low-alloy or high-strength steel, most of it manufactured in England. The low-alloy steel, which was *rolled*—a term referring to the process of rolling steel ingots into plate steel—was to be of fusion welding quality,

Star, sway or wind bracing, and lateral diagonals keep the structure rigid against the ravages of wind and the rapid acceleration and braking of vehicles. PHOTO BILL CUNNINGHAM, *THE PROVINCE*

Stiff-legged derricks, known as travellers, were used to boom the steel down to the raising gang. PHOTO P. STANNARD, DOMINION BRIDGE CO. LTD.

that being a process by which two parts can be joined to form a permanent bond. The steel specified for the job was classified as ASTM 242 or equal, the British equivalent being B.S. (British Standard) 968. As much as approximately 20,000 tonnes of steel would be required to build the bridge, the equivalent of about 20 percent of the province's annual steel consumption.

After this structural steel was fabricated into various bridge components at the plant, it was delivered to the north end of the bridge by rail or truck depending upon the size of the member, where it was offloaded or, in ironworking terms, shaken out at Span 1 by a yard derrick known as Traveller No. 3, one of three "travellers" working the site. The term traveller refers to a crane, or stiff-legged derrick, that pulls itself along the ground when it needs to move. The stiff legs angling down to the bridge deck from the top of the mast were known as moon-beams.

Traveller No. 1 and No. 2, which were on the bridge, were anchored to steel floor stringers, and when they were required to move, they were jacked up, steel blocking removed, and they were then lowered onto steel rollers. Tentacle-like cables, known as tackle, were drawn out ahead of the machine and anchored to whatever piece of steel could handle the load, usually a floor beam. Another engine aboard the traveller, an International four-cylinder gas engine known as a donkey, then winched the cables in, slowly advancing the unit forward in usually jerky movements as it rumbled over the elevated jumping plates covering joints in the floor stringers. Each move took about 2.5 hours. The four jumping plates, each weighing 63.6 kilograms, had to be lugged ahead of the traveller by hand to cover gaps between the floor stringers and the floor beams.

Traveller No. 3 lifted the steel in the order it was required at the front end up onto two small flat-deck rail cars called bogies located on the west side of the bridge deck. There, it was met by Mel Alexander, foreman of the backend crew, whose gang consisted of another six or seven men. They would block the steel on the bogies to prevent it from shifting during transport to the front end. Other members of Mel's crew were responsible for attaching lifting lugs to larger members which they did in the yard prior to the piece being hoisted up to the bridge deck, and still others did nothing but scale rust from the many steel gussets required to erect each piece.

Every steel member that was boomed up had painted on its side the contract number of the job, i.e. *S.3703E*, in addition to a number corresponding to the accompanying "erection procedure drawing," detailing the erection method and the type of equipment required to install it. These drawings showed the bridge as it would ultimately look. Attached were another set of instructions called "erection diagrams," showing "the location of all the members by marks or numbers or both, and the order in which these members are to be placed on the permanent structure."[43] Essentially, each bridge member was designed, drawn, detailed and fabricated according to shop detail drawings—instructions from the engineers to the fabrication shop—and then marked with a number to correspond with the various drawings and diagrams that arrived with the piece on the bridge.

The marks on each piece were called "shipping marks" or "erection marks." If the piece was to be installed on either the left- or right-hand side of the bridge, it would be marked with an *L* or an *R* together with its identification number. If it was a chord, it would be marked with a *BC* for bottom chord or a *TC* for top chord. In the field, the men often applied different terms to various pieces; for example, bottom chords were referred to as lower chords and top chords as upper chords. If several pieces of steel were detailed on one plan, they would be differentiated with an *A*, *B* or *C*. It would be almost impossible, in fact, with such detailed instructions, to place a piece of steel in the wrong location, backwards or upside down.

The bogies were pushed by a small locomotive run and maintained by a member of the International Union of Operating Engineers Local 115, Gordon MacLean. In a DB photo taken of the men of the Second Narrows Bridge, MacLean is the ample man with the stoic look sitting in the front row on a wooden bolt keg off to one side. If the photo had been in colour, one would have seen that he was blond and blue-eyed. The fact that he was an operating engineer is apparent from his dress—greasy coveralls and a cocky engineer's cap—in contrast to the hard hats worn by the ironworkers around him. Operating engineers figured they didn't need lids because they were always in their cabs, but High Carpenter made them wear their hard hats when they were off their machines. MacLean was fifty-three and had already spent a lifetime messing with grease and hot engine oil: first, serving his country during World War II in the tropical stokehold of a Canadian naval escort vessel plying the North Atlantic, and later running logging shays on the narrow gauge tracks winding through the dripping coastal forests of Northern Vancouver Island. Now he was on the bridge.

Big engines were his life, and he was perfectly content to sit in the cab of the 25.2-tonne locie, listening to the purr of the big gas engine as it shunted steel out to the front end of the bridge on its standard gauge track. On a typical haul, the locie passed Traveller No. 2, a 25.2-tonne stiff-legged derrick that assembled and disassembled the erection harnesses used to join two steel spans during the erection process. It then carried on out to the front end of the bridge where Charlie "Grumpy" Geisser (pronounced "Giser"), another war vet, operated Traveller

The men who hung the steel of the new Second Narrows Bridge. From left to right, front row: Fred Barriball, Lou Lessard, Gordon MacLean, Vern Siedelman, Charlie Geisser, Palmer Enger, Jim English, Bill Stroud, Stan Marshall, Joe Petriska, Charlie Moore, Bill Wright. Second row: Norm Atkinson, Joe Chrusch, Larry Cameron, Ted Barkhouse, John Wright, Bill McDonald, Jim Marshall, Doug Bradford, Bill Adair, Dennis Gladstone, Jack Thompson, Rod Smith, Jim Pratt, Ted Twining, Eric Guttman,

Frank Hicklenton, Bill Lasko Sr. Back row: Bobby Robertson, Ron McDonald, Mel Alexander, Nick Scherst, Stan Gartley, Lloyd McAtee, Ed Raglan, Vince Spinder, Gary Poirier, Sam Rouegg, Walter Carter, unknown, Bern White, Murray McDonald, Paul Yurchuk, "High" Carpenter, George Ferguson, Fred Leenstra, Bill Moore, Colin Glendinning, Alan MacPherson, Cliff Tate, Dick Mayo, Kevin Duggan. DOMINION BRIDGE CO. LTD.

No. 1, a colossal 111.1-tonne stiff-legged derrick with a 41.1-metre boom (boom length varied between 26.0 and 41.1 metres depending upon weight of the lift). It lifted the huge steel members, some weighing as much as 96.5 tonnes, from the bogies and lowered them to the raising gang waiting on their wooden floats below. Geisser was a bit of a celebrity in that his name had been drawn three years earlier to join the dignitaries snipping the ribbon on the new Granville Street Bridge. He had even been presented with the silver-plated scissors after the event.

Geisser, similarly garbed to MacLean, appears in the same DB photo two men to the right. When working, he sat on a stool in a wooden shed built around the traveller's controls consisting of about a dozen levers plus foot pedals that he manipulated to spool, un-spool and brake the three main drums for the main line, auxiliary line and boom falls, as well as the boom-swinging drum. The traveller's hoist was powered by a 160-horsepower Cummins diesel with a Twin Disk torque converter. Cables ran from the drums over a tall mast supported by two stiff legs anchored to permanent bridge stringers. From there they ran along to the top of the boom before dropping to the load fall block with its lifting hooks. Another line, called a runner line, ran across the boom and over its own pulley, dropping to a large steel ball called a headache ball, with a hook that overhauled the line when it was needed to lift smaller members. This was the ball that the men were often seen riding around on. The whole boom assembly turned on a giant wheel called a bull wheel. Behind the shed, barrels of diesel and gas were stored to fuel the traveller's main engine and the crawling donkey engine.

Ironworkers called the working end of a bridge the front end, which to the uninitiated would appear to be a chaotic place. Alive with smoke, noise, dust and movement, it was the movement that took some getting used to. With the arrival of the locie and the slewing action of the traveller, the bridge bounced and jostled. Ironworkers call this "live," and know it to be a healthy sign, for rigid steel can snap. To the men who worked the front end, comfort came from knowing that a well-rehearsed system of checks, balances and procedures would shield them from most of the risk. It also required that each man know and do his job, the unwritten skill set of which demanded that lightning reflexes, instinct, courage and brute force be intrinsic to this industrial high-wire act that either made these men good at their jobs or down the road. It also didn't hurt to have a mechanical aptitude, be in good shape and have a solid work ethic. There was no room for error at the sharp end. Sloppy work made for unsafe conditions, which put men's lives at risk—the job was dangerous enough without adding to it.

Three gangs formed the core of men at the front end, with two of the gangs having roughly four members and one having eight. Each gang had a foreman and

The "locie," the small gas locomotive used to haul steel out to the front end, is lowered onto its tracks. It hauled two small flatcars called "bogies." COURTESY BOB DOLPHIN

often an apprentice. In addition to the members of the hooking-up gang, raising gang and bolting-up gang, there was MacLean and Geisser running the locie and traveller respectively, as well as a signalman and a job superintendent among a myriad of others.

The job super in this case was Jim English. He had initially joined the industry after surviving two severe wounds with Vancouver's Seaforth Highlanders in Sicily during the war. Upon returning home to Prince George, he was invited by his uncle, who lived in Vancouver, to come down to look for work. He soon found a job with DB, and had been with them for four years when he was assigned the responsibility of conveying Murray McDonald's directions to the various gang foremen—to keep the job on track and on time. English was a no-nonsense, rough sort of guy, a cowboy on leave, with an easy sense of humour but whose body language cautioned *don't mess with me*. He was tough but respected, two essential qualities of a boss in charge of a bunch of other hard cases. "A man of this calibre has to be this way," says ironworker Gary Poirier, matter-of-factly.

The hooking-up gang, the first gang at the front end to touch the steel, had three members excluding the foreman, Lucien Lessard: Jack Thompson, Bill Wright and Alan Stewart, an apprentice. Stewart, twenty-eight years old, was a member of another family legacy; his brother Wayne and his dad Sid were both ironworkers. Having a family member in the union was a benefit in that you were automatically shunted to the head of the apprenticeship line. Lessard's hooking-up crew would set the chokers or slings required to prepare the steel to be lifted off the bogies by the traveller. In some instances, lifting lugs had already been bolted to larger members by the time it reached the front end, and always, tag

or tugger lines were attached so that the men could guide the steel once it became airborne to control its pitch and yaw. A wild piece of iron was an accident waiting to happen.

But the iron needn't be in the air for it to be dangerous. When Gary Poirier was an apprentice, there was

"Big" Jim English had been with Dominion Bridge for four years before he was assigned to the Second Narrows Bridge project as job superintendent. PHOTO BILL CUNNINGHAM, *THE PROVINCE*

a time when he, Mel Alexander and a journeyman were attaching a lifting lug directly to a piece of steel, and the drill bit got caught in one of the eight holes required to bolt down the plate. The journeyman threw the machine in reverse, hoping to back off the bit, but instead, the machine began to gyrate uncontrollably, knocking all three men off the steel and the bogie where they were working. Alexander and the journeyman were thrown east toward the centre of the bridge, but Poirier was thrown to the west, catapulting over the bogie to land a scant metre from the edge. Poirier remembers it as if it were yesterday.

"Where's the kid?" shouted Alexander worriedly as he struggled to rise.

"The kid's over here," replied a bruised and somewhat bemused Poirier as he poked his head up over the bogie to confront the errant drill. "Danger comes when you least expect it."

Communication was vital at the front end where the steel was usually below or out of the line of Geisser's sight. Frank Hicklenton, who had been John Prescott's batman during the war, was the signalman. He acted as Geisser's eyes using a series of hand signals (radios were not in use then). Standing near the traveller, he would receive a signal from Lessard that the lifting chokers or slings were set and that the load was ready to hoist. He would then pass this message to Geisser in the shed. The black cloud of soot coughing from the Cummins diesel, flipping the rain cap on the exhaust stack as the engine roared to life, confirmed that Geisser had received the message. He hoisted the load just enough to determine whether it was on-centre or not, but Lessard's crew prided themselves on finding the mark first time. If it were off, however, Lessard would signal to Hicklenton who would signal to Geisser to set it down to be repositioned. Once the load was again ready, it was picked up and boomed down to the raising gang below.

Signalling the steel being boomed down to the raising gang waiting on their wooden plank floats. DOMINION BRIDGE CO. LTD.

That gang had four members, again excluding Lessard, who also served as their "pusher," or foreman: Bill Stroud, Joe Chrusch, Colin Glendinning and Norm Atkinson. The raising gang, a.k.a. connecting gang or erection gang, represented the pinnacle of the ironworking profession and the men who had this job knew it. They were cocky, some would say arrogant, but this was expected of the men in this position. They also had trust, and the bonds that cemented a raising gang not only protected their lives but also grew into friendships that sealed a pact: you watch my back and I'll watch yours. These men were at the front of the front end and they were damn well proud of it.

Lou Lessard was a young, thickly accented ironworker from Quebec. Sinewy, as honest as he was opinionated, his chiselled features reflecting the rigours of his profession, he fell in love with the idea of walking high steel in 1951 when he first saw men doing it at the DB plant at Lachine, Quebec. It was a short jump from romance to bridgework. Lessard's prior bridge experience consisted of hanging steel on the Angus L. MacDonald Bridge between Halifax and Dartmouth, a suspension bridge much like the Lions Gate Bridge. Deciding that he wanted to be in Vancouver, he wrote to John Prescott, construction manager with DB at the time, to inquire about work there. Prescott wrote back, telling him to look him up when he was out this way, and two weeks later Lessard showed up ready to be put to work. Prescott remembered the conversation.

"Well, it's been kind of slow, you know, and you're going to have to join the union," Prescott cautioned, thinking that the fifty to sixty men looking for work ahead of him would slow him down a bit. Two days later Lessard was back.

"Well, did you join?" Prescott asked, cautiously.

"Yeah," answered Lessard.

"How did you manage that?" Prescott asked, incredulously.

"I slipped the business agent a hundred bucks." Lessard grinned.

With that sort of ingenuity, Prescott felt he couldn't miss. Within the week, Lessard was on the job in the sticks, but his confidence and ability soon escalated him to the raising gang foreman of the new Second Narrows Bridge, where he would become one of the few salaried employees of the giant steel erector. It doesn't get much better than that for a connector.

Norm Atkinson was the epitome of a connector, having learned some very transferable skills aboard a number of Canadian Navy vessels in the North Atlantic during the war. A leading seaman, top rigger and then petty officer aboard the heavy destroyer, HMCS *Ottawa*, he was responsible for all the ropes, cables and splicing aboard ship. It was a natural progression from ship to shore. Sailors learned how to deal with heights as well as how to rig a ship—essential skills for a connector.

"If you can't deal with heights you have no business at the front end . . . You should be able to work and climb at the same time," he says.[44] Don Heron remembers that Atkinson was "catty," or nimble, on the iron but the humble ironworker would only concede that "It was our office, and we just went to work every day."

Balance follows confidence, and soon, walking a fifteen-centimetre flange fifty metres above the ground was no more bother than walking on the sidewalk, except that you had to keep your wits

Norm Atkinson, connector – "If you're afraid of heights you have no business at the front end." COURTESY LOCAL 97 about you; tripping was a short journey to hell. "Look where you're going, not where you're at,"[45] warns Ron Rollins, head of the ironworking department at the BC Institute of Technology.

Bill Stroud not only had no fear of heights, but he was drawn to high places, having been a high rigger in the logging industry prior to becoming an ironworker. He thought nothing of scaling a spar tree and sitting on top thirty metres or more above the ground, taking in the view.

Stroud's partner Joe Chrusch, in contrast, hardly fit the bill of a typical connector, or even an ironworker for that matter. This was not because he lacked ability—he was as skilled as the next guy—but just because of who he was. You see, Chrusch was a bit of a renaissance man. Brought up on a farm in Lac du Bonnet in rural Manitoba, he initially made his living in construction and logging camps. But that was before driving home over a frozen lake one spring day nearly cost him his life. His car hit a rotten bit of ice, plunging it and a bewildered Chrusch into a netherworld of certain oblivion. While he was contemplating his predicament, however, the car hit bottom. Before the mud could settle, Chrusch was already making a slow ascent to the halo of light above. His guardian angel was awake that day. Ironworking seemed tame after that adventure.

But this was not what made Chrusch different. In addition to his love of heights, and being described as a daredevil by his daughter Diane, he was gifted at the softer skills of life. He was a talented tailor and cook. Who knew? Not only was he teaching his daughter Maureen how to knit a sweater with a bridge design on it, but he was making her a black skirt for her choir assembly as well.

"A man's got to be very comfortable in his own skin to take on tasks like that," reflected Maureen. It was this sort of confidence that made Chrusch an invaluable member of the raising gang.

As Geisser boomed the steel down below the forward bridge horizon, Hicklenton turned to focus on Lessard who was watching the steel and would give Hicklenton the hand directions to

Joe Chrusch was not only a valuable member of the raising gang, but he was also a bit of a renaissance man in that he was a talented tailor and cook as well. COURTESY CHRUSCH FAMILY

relay on to Geisser. Slow, down, down, down, over, down, slow, his hands would flutter. Jogging the steel down in imperceptible increments, Geisser reacted to Hicklenton's caution as Glendinning reached up for the tugger line that came within his grasp, and sucked in the massive steel beam by leaning into it with all his weight.

The raising gang had a tool kit that was specific to their needs. Weighing as much as twenty-three to twenty-seven kilograms, depending upon how many bolts filled their pouches, the kit was attached to their waists by a quick-release belt, an oh-so-important feature in case they took a swan dive into the inlet. Slamming into the water from fifty or more metres up is much like hitting concrete. If they survived the impact, the weight of the belt was more than the average SCUBA diver wore, and would quickly carry them to the bottom if they couldn't get rid of it. Among other things, each kit generally had two spud wrenches, tapered at

Lou Lessard supervising the connection of an upper east-side chord. DOMINION BRIDGE CO. LTD.

95

the handle end, one or two twenty-centimetre double-tapered barrel drift pins for aligning holes, a long connecting bar tapered at one end with a flange at the other, and an approximate four-kilogram sledgehammer called a beater to pound in the drift pins or to get the attention of any uncooperative piece of steel. One or two pouches contained the bolts they would need for a particular connection.

As the steel edged toward the join, Glendinning and Chrusch wrestled it into place while Atkinson stuck the tapered end of his spud wrench through holes in the beam and the permanent steel to hold them together until his drift pins could align and further secure the connection. Atkinson wore his watch upside down to prevent the stem from digging into his wrist as he hammered the pins home with a rapid-fire set of ringing, crab-like blows. He would leave them for the bolting-up gang to remove. A dozen bolts on either side were quickly inserted and buzzed tight with an impact wrench. All of this activity carried out on two precarious wooden-plank floats hung on either side of the steel.

As the chatter of the impact wrench reached Geisser in the traveller's shed, he sat back momentarily to admire the view before the raising gang released the cables from the steel. On Hicklenton's signal, the Cummins diesel leapt to life as Geisser lifted the boom and swung it back to the locie waiting with a matching piece of iron for the left side of the bridge. Marching into the inlet; left, right, left, right, left, right, life was grand at the front end on a clear day.

Following quickly on the heels of the raising gang was the bolting-up gang. The eight men including the foreman, Rod Smith, were Gordon Schmidt, Ted Barkhouse, William Adair, Dennis Gladstone, Carl Holmstrom, Donald Gardiner and Donald Mitchell. Four worked the east side and four the west. Perched on the raising gang's floats, they worked quickly to secure the connection with their impact wrenches.

Tightening bolts to most of us would appear to be an uncomplicated task, but to an engineer it is a complex science involving metal elasticity, tension, and loads. Although there are a few standard bolt-tightening methods, Angus McLachlan, chief engineer of DB's Pacific Division, wrote to Swan, Wooster & Partners in November 1957 to propose that the company use the "turn of nut method," which they had recently successfully employed on the Peace River Bridge at Taylor, BC, and on the Nine Mile Canyon Bridge across the Fraser Canyon. This method, which is now called angle-controlled tightening, was only introduced after World War II for application with power wrenches. It calls for the bolt to be tightened to a specific angle, the only downside being that precision is required, and the bolt can only sustain a few reapplications before breaking.

To check the tension on the bolts, Frederick Leenstra, a Swan-Wooster

Checking the tension on the bolts with the torque wrench, a metre-long tool with a gauge and a light at one end to signal that the correct torque has been reached.
PHOTO BILL CUNNINGHAM, *THE PROVINCE*

employee, and Gary Poirier used a torque wrench to ensure that they were tightened to specification. The wrench, which was almost a metre in length, had a gauge at the business end as well as a light to signal that the right torque, nine hundred-foot pounds, had been reached. Poirier was also the go-to guy for the bolting-up crew and kept them charged with bolts and whatever else they needed. Born in New Westminster in 1940 to a French Canadian father and Thompson River Salish mother, he didn't see much of his dad, who divorced his mom soon after returning from the war. She then moved Gary and his brothers, Neil—who also became an ironworker—and Ian, into Vancouver. Gary quit school at age fifteen and began a series of jobs in mills, on the railroad and in logging camps before developing a skill at reinforcing iron, which ultimately led him to join Local 97. After his apprenticeship, he transferred to structural steel and never looked back.

"As soon as I came out of that apprentice school, after being on rebar, I never went back to rebar, I stayed with structural steel. I worked at Nine Mile Canyon Bridge—it was a cantilevered bridge also. So I got knowledge with leaning out with the bridge itself. We never had safety lines whatsoever and you had yourself and that was the challenge, and you had to work too, see," he recalled. Bridgework was his life, and as with most ironworkers, once they had tasted work on a bridge it was hard to go back to anything else. "Bridgework gets into your blood," Poirier reflected, "and it never lets you go." Perhaps it is so appealing simply because of the nature of the structure, the risk, its majesty, or the fact that few, especially large bridges, are ever built.

On the heels of the ironworkers came the painters who worked for Boshard & Son, the painting contractor. Each piece of steel that left the plant had already been coated with a red lead primer to protect it from the elements, but it needed a final coat, which Boshard's men were applying in the field. Curiously, the primer began to bubble a couple of months after erection. Angus McLachlan asked

Gary Poirier, right, and his brother, Neil; both ironworkers. COURTESY GARY POIRIER

Byron Maine, DB's paint inspector, to investigate. "What was happening was the paint started to blister and pop off . . . it was due to the steel coming over from England on deck . . . and you've got saltwater and brine impregnated into the steel and it should have been washed off with fresh water, but they didn't do it. Consequently, the paint started a reaction between the salt brine, the carbon steel and the red lead primer. I made sure that the guys scraped it right down to bare metal and wire brushed it before they put more paint on," Maine recalled.

By early March 1958, three steel spans stretched south and a fourth was partially cantilevered toward Pier 14, the north anchor pier. Each span was a complex confusion of steel, each member vital to the integrity of the whole. A typical span, which stretched 86 metres between two concrete piers, comprised two trusses, one on the east side of the bridge and one on the west side. A truss is made up of a top chord—a chord being a massive fabricated steel box large enough for a man to crawl through—and a bottom chord. Connecting the two were vertical structures called plumb posts and angled structures called diagonals, the latter two making up what is called the web of the truss. At each joint there were steel plates called gussets securing the connection.

Joining the chord to its mate on the other side of the bridge are three bracing systems: a top lateral brace running horizontally (east to west) between the two top chords and a bottom lateral brace running horizontally between the two bottom chords. Lateral diagonals, offering additional support, fan out from the centre of each bottom lateral brace to land at either end of its neighbouring brace. The third bracing system runs diagonally between the top of the verticals of the east chord and the bottom of the verticals of the west chord, and vice versa, to form what is referred to as star, sway or wind bracing. Essentially, the lateral diagonals and the star bracing guard against stresses from the wind and the rapid braking and acceleration of vehicles, thus making the structure rigid.

Riding over the top chords is the floor system consisting of floor beams running east to west and stringers over them running north to south along the length of the bridge. There is a floor beam at each panel point—a panel point being described as the joint where a plumb post and diagonal meets a chord. The area between each panel point is referred to as a panel, which represents twelve metres of iron or about two days' work. Progress was marked by panel points. The roadbed would eventually consist of a 13.9-centimetre-thick layer of lightweight reinforced concrete formed from shale aggregate poured over the stringers, with a 5.08-centimetre layer of asphalt spread over that.

To build Span 4 it was necessary that Span 3 be complete, as it would provide the anchor to allow Span 4 to be cantilevered. Span 4 therefore had to be

Mock-up of the bridge prepared for the royal commission July 31, 1958. OTTO
LANDAUER OF LEONARD FRANK PHOTOS, JEWISH MUSEUM ARCHIVES LF-35627

tethered to Span 3. In a completed bridge, each span is independent of the other, relying only on the concrete columns for support, but during erection the steel of a cantilevered span is attached to its neighbour. Essentially Span 3, stretching from Pier 12 to 13, and the cantilevered Span 4, became one continuous unit with Pier 13 acting as the fulcrum for the two spans.

Cantilevering is the most dangerous part of the erection process where all sorts of forces come into play. As the weight of the cantilevered Span 4 wants to drag it down, the north end of Span 3 wants to spring up, much like a teeter tot-ter. Both actions must be controlled. To prevent Span 4 from dropping—in other words to resist the tension being applied to it—an erection harness was installed over Pier 13 to join the two spans, which also permitted the cantilevered span to be controlled. In other words, the harness allowed Span 4 to be jacked up or "cocked up" so that it would arrive high at its landing point. The span could then be lowered onto Pier 14 when it was ready to do so. The erection harness, which was similar in design and look to a truss over a beam bridge, was none other than that used by Colonel Swan on the Pattullo Bridge twenty years earlier, only the struts had been beefed up for the current application. He called it "an old friend of mine."

To prevent the north end of Span 3 from springing up, it was held in place with tie-down bars that descended twelve metres into wells set in the concrete columns of Pier 12. Using hydraulic jacks, two on the east side and two on the west side for a total of 227 tonnes of resistance per side, the span could be raised or lowered depending upon the need. Tie-down bars are especially important when the weight of the cantilevered span being erected is equal to or greater than the weight of the anchor span; however, if the anchor span is much larger than the span being erected, tie-down bars are not always considered necessary. A further

Erection harness over Pier 11 to support the cantilevered Span 2 while under construction. OTTO LANDAUER OF LEONARD FRANK PHOTOS LF-35120, COURTESY JIM ENGLISH

The bridge rests on the bent N4 falsework, only a few days from catastrophe.
PHOTO BILL CUNNINGHAM, *THE PROVINCE*

measure to prevent the erected span from dropping was to install kicker blocks or plates, weighing about half a tonne each, between the lower chords of the two spans to resist the compression that wanted to force them together. In addition to the weight of the steel, the erected span also had to bear the weight of the traveller plus the locie and its two bogies, for a total additional weight of 120 tonnes. So, to eliminate any unnecessary loading while Span 4 was being built, only necessary floor beams and stringers were installed until after it was landed on Pier 14 and could bear the additional weight.

Once the erected Span 4 was poised over Pier 14, the north end of Span 3 was jacked up to tip the south end of Span 4 onto its pier, Pier 13 acting as the balancing point. When the south end of Span 4 was settled onto Pier 14, the erection harness was dismantled using Traveller No. 2, and the north end of Span 3 jacked down to its resting position. Both spans then became self-supporting on their respective piers. Although independent, they were not static. The steel of a bridge is a living thing, expanding with the heat of day and contracting with the chill of night. To accommodate this movement, each steel span is anchored at its north end by a fixed shoe with slotted holes fastened with 4.4-centimetre anchor bolts, but rests on a nest of steel rollers at its south end so that it can roll out during the day and roll back at night. As well, on the bridge deck, every thirty-six metres metal expansion plates were installed to permit the road surface to expand and contract without damage.

The next span, Span 5, was formally referred to as the north anchor span. It would carry the steel from the north anchor pier, Pier 14, south to Pier 15, the north main pier, which was not much more than a pedestal, from where the steel would launch across the inlet. Compared to the previous four spans, this span would be a lengthy 141.7 metres and would require a different method to support it during its construction. The erected span would cantilever approximately fifty-two metres from Pier 14 before coming to rest on what is termed *falsework*, or false bent-work. There would be two sets of falsework required for Span 5.

Falsework is roughly described as a steel frame built to support a structure during construction. The term "false" relates to its impermanence as it was removed once the span reached the next pier and became self-supporting, while the term *bent* refers to a substructure with columns and caps. The base of the first falsework of Span 5 was supported on two cement-filled, steel-pipe pile nests—one nest directly below the east bridge truss and a similar one on the western side—driven into the floor of the inlet by Greenlees Pile Driving. The base itself consisted of two tiers of I-beams on each pile nest called a lower and upper grillage. The lower grillage consisted of four short beams laid side by side and oriented north to south—four to each pile nest. The upper grillage consisted of four long beams, also side by side, running east to west across both sets of lower grillage.

Placed on top of each end of the upper grillage was a 14.6-centimetre-thick steel plate shop-welded to a leg that rose up to another steel plate or cap that was attached directly to the bottom chord of the bridge truss, one leg on the west chord and one leg on the east chord. Both the tops and bottoms of the legs were socketed to permit the assembly to rock back and forth with the expansion and contraction of the steel. Lateral and diagonal bracing between the two legs of the bent locked them together to form a single, rigid steel support. Although fabricated at the plant specifically for this use, the steel of the falsework would later be dismantled to form part of the permanent bridge.

Until it reached its first set of falsework, Span 5 would be tethered to Span 4 with tie plates, four plates connecting the top chords of each truss. An erection harness was considered unnecessary. To use tie plates, when the top chords of Span 4 were designed, they were fabricated with longer webs than normal. The web of a chord is simply the sides of the boxed structure. The tie plates were bolted to the top-chord webs of Span 4 with ten bolts and connected to the top-chord webs of Span 5 with 31.8-centimetre-diameter pins, which held the cantilevered structure until it could land on its falsework. Then, the erection would be cantilevered off the falsework to either land on another set of falsework, which it was designed to do, or onto a pier. Kicker blocks were also installed between the lower chords of the two spans to resist the compression there.

It was one of John McKibbin's responsibilities to design the Span 5 falsework, Bent N4, that was to be approved by his boss Murray McDonald. When McDonald and McKibbin were not on the bridge managing the erection or taking measurements, they were domiciled in a small mobile office on the north bridge approach. It was important that they be close to the bridge and it was in that office, on June 29, 1957—almost a full year ahead of the bent's erection—that McKibbin completed the grillage calculations with a slide rule. He used the Canadian Standards Association guide as a model despite the fact the American Association of State Highway Officials (AASHO) guide was being followed to build the bridge. It would be discovered later that this had its own set of problems. McKibbin used a standard DB worksheet which he entitled, "Design of Caps and Distributing Beams Using 36 WF160 Beams between Pairs of Bents." The 36 WF referred to the 36-inch depth of the wide flange I-beams while the 160 referred to the beam's weight per foot.

McDonald reviewed McKibbin's neat figures and marked *OK* beside his work. Based on McKibbin's calculations, the upper grillage would not require web stiffeners (steel supports running vertically between the flanges of the I-beams) because the load-bearing capacity of the beams was considered adequate to support the approximate 2800 kilopounds bearing down upon it. Unbeknownst to McDonald, however, McKibbin had made several errors, including two significant dimensional errors. The first of these dimensional errors was in calculating the need for web stiffeners. To arrive at that calculation, the thickness of the web, or middle part of the I-beam, was required. McKibbin took the 2.54-centimetres flange thickness by mistake, rather than the web thickness, which was only 1.65 centimetres. Using his figure and the load that would be applied to the four upper grillage beams, he arrived at a value of 14 kilopounds per square inch when in fact it was more accurately 21.7 kilopounds per square inch.

Below that calculation was one for shear force and from that was calculated shear stress, a term referring to a "stress tending to produce or to resist a shear"[46]—the word "shear" simply meaning to distort or fracture. This latter calculation sought to determine "the shear force divided by the cross-sectional area in square inches of the web of the beam,"[47] the area referring to the part of the beam that resists the shear. The web area for 36 WF 160 beams was 23.51 square inches as tabled in the *Handbook of Steel Construction* published by the Canadian Institute of Steel Construction. Beside that column was another list of figures providing the gross area of the whole beam including the top and bottom flanges plus the web. This factor, which was precalculated at 47.09 square inches, double that of the web area, was the one that McKibbin erroneously selected for his calculation. The result was that the shear stress would be only 6 kilopounds per square inch, when in fact it was more than double that at 14.3 kilopounds per square inch.

Based on these calculations, McKibbin determined that steel web stiffeners would not be required,[48] despite on the same day finding stiffeners and diaphragms (supporting steel laid across the top flanges of the I-beams) necessary for the lower grillage I-beams. Instead of diaphragms, 2.54-centimetre milled rods with turnbuckles and hooked ends were installed at each end of the upper grillage to cinch it together. With respect to the lower grillage, the first calculation sheet that McKibbin had prepared was marked *NBG* (no bloody good) in the bottom right-hand corner by McDonald, causing McKibbin to rework the calculations on another sheet. Had McKibbin arrived at the correct figures for the upper grillage, he would have found that not using web stiffeners would have contravened the AASHO standards, which were being followed to build the bridge, and which permitted a maximum of only 8.25 kilopounds per square inch for rolled steel without stiffeners. Even adding a standard 25 percent to 33 percent buffer for erection stress, including full erection wind stress, McKibbin's figure of 6 kilopounds per square inch was still within the range of permissible stress without stiffeners, according to AASHO.

The design of the whole falsework assembly had also been prepared by McKibbin and checked by McDonald, and had then been forwarded to DB's erection department for preparation of the drawings to be passed on to the detail shop

Critical design sheet prepared by John McKibbin and checked by Murray McDonald a year before the collapse. BC ARCHIVES GR-1250

prior to being sent to the plant for fabrication. Although the design sheet showed no stiffeners or diaphragms in the upper grillage, nobody in those departments, largely made up of draughtsmen, would question the authority of the erection engineer and therefore would prepare the bent as presented on the plan. Although the erection manager was responsible for "ensuring that all erection work conforms to the applicable codes, standards and specifications,"[49] he couldn't possibly have been expected to check the hundreds of calculation sheets that flowed across his desk. And there was no need to forward the sheet with the shear calculations to the erection department as no fabrication was required; the I-beams for this portion of the falsework were already on hand. This calculation sheet therefore remained on the bridge.

The mistakes that were made manifested themselves into a series of flawed assumptions that would prove especially critical. These calculations supported McDonald and McKibbin's decision to use floor stringers for the upper grillage, stringers that they may have planned to use later in the permanent bridge deck. Individual I-beams were not that expensive—perhaps about $1,000 to $2,000 each—but there was always a responsibility for an erection engineer to be frugal and "cost control" was one of McDonald's responsibilities. Using floor stringers was an acceptable practice as long as the stringers being used would support the load and providing they were not damaged for later use. Welding was therefore out of the question, but any bolting was considered acceptable as long as the bolts were replaced in the stringer after the grillage was dismantled and before the stringer was reused.

Since the calculations suggested that the stringers would be adequate for the job, instead of steel web stiffeners and steel diaphragms, 30.5-square-centimetre wooden blocks, 90 centimetres long, were inserted vertically between the flanges of the I-beams approximately one metre from the centre of the load. These blocks, held in place by steel straps, would provide negligible support, for wood has little crushing strength relative to steel.

There were also problems with the stringers that had nothing to do with the calculations, but nevertheless, had all the earmarks of trouble. Although rolled steel members are never perfect, the subject stringers were of uneven depth; the flanges were not parallel and were slightly off-perpendicular with respect to their webs; the widths of the left and right sides of the flanges were unequal; the top flange of one beam had a small off-centre depression, which reduced its load-bearing capacity; and the web of one beam had a slight curvature. The different stringer heights were especially worrying, as the highest would bear more of the load, but not so disquieting as the slight curvature of one of the webs. To absorb the height differences, the upper grillage was sandwiched between sheets of 19.05-millimetre plywood called "softeners," which was a convenient method to even out the load.

The year that transpired between McKibbin and McDonald making and reviewing the falsework and grillage calculations was busy, and the last thing on anybody's mind was to review calculation sheets that had already been checked and filed. The calculations were on record, a competent engineer had checked them, and the support had been designed and detailed. No red flags had been raised, and for McDonald and McKibbin there was no question that it would not be sound. The critical calculation sheet therefore remained hidden in its folder in the engineer's filing cabinet—a sleeping catastrophe. Someone, however, revisited the sheet between its calculation in June 1957 and the collapse a year later, and what they discovered, and what they failed to do with that discovery, would astound the royal commission that investigated the collapse.

Engineers hold themselves to a high standard. Mistakes made by them may have grave consequences, and it is for this reason that there is a time-honoured system of checks and balances to prevent this from happening. As far as McDonald and McKibbin were concerned, there was no cavalier attitude on the bridge, nor was there any laziness; there was, however, an inexperienced associate engineer who was, by his own admission, a little overawed by the responsibility being handed him. There was also a harried erection engineer who was performing three different functions. But there was something else as well, and perhaps that was even more alarming; there was no clear procedure for dealing with falseworks. But falseworks were common to bridge construction, and in the previous twenty-five years nobody could remember DB having a failure in that area.

Few engineering firms at the time checked falsework calculations and designs, even though the Second Narrows Bridge contract between the BC Toll Highways and Bridges Authority and DB called for it (Swan-Wooster only checked falseworks to the point that they thought they needed checking). Interestingly, this clause had only been introduced in 1954 during construction of the Agassiz-Rosedale Bridge, and the Second Narrows Bridge was perhaps only the next incidence of its occurrence. Page 19, paragraph 2-2-3 of the contract, stated that:

> The contractor shall furnish, construct and subsequently remove all falsework required for the erection of the steel work. Falsework shall be properly designed and substantially constructed and maintained for the loads which shall come upon it. The contractor shall submit to

the engineer plans showing the falsework he proposes to use to enable the engineer to satisfy himself that the falsework proposed to be used complies with the requirements of this specification. Approval of the contractor's plans shall not be considered as relieving the contractor of any responsibility.[50]

Despite the contract calling for it, that McDonald and McKibbin did not submit the falsework plans and calculations to Swan, Wooster & Partners was not unusual. Most engineers considered this type of construction fairly routine and did not insist on seeing falsework plans. In fact, the AASHO had no requirement to submit falsework plans whatsoever, as the falsework was not considered a permanent part of the bridge; it was, in fact, more of a piece of equipment. Even the Canadian Standards Association's *Specification for Steel Highway Bridge Manual* for 1952, makes little mention of falseworks other than the contractor shall be responsible for the correctness of his drawings, falseworks were to be furnished and constructed by the contractor, and it was the contractor's responsibility to remove any falseworks and clean up the site after.

John Prescott remembered that McKibbin and McDonald had both initialled the bent-work design, "and then it had gone to the engineering department, Angus McLachlan's department, and he had assigned an engineer to the job by the name of Cliff Paul. Paul worked for McLachlan. There's some dispute there. Paul said he was never assigned to the job and McLachlan said he was assigned to the job . . . They don't deny looking at it but they deny any responsibility for checking it."[51] It was an argument, however, that could have no winners at that point, whether settled or not, which it never was. Besides, the critical DB worksheet with the faulty grillage calculations had likely never left the engineering office on the north bridge approach and therefore could not have been known to either McLachlan or Paul.

On April 1, 1958, the N4 falsework was erected under Span 5, and the partially erected span jacked down onto it. Work continued until April 25 when all erection activity on the bridge ceased for the next six weeks. DB had run out of steel due to a miss-shipment from England. Although the steel would arrive in Vancouver within a couple of weeks, it would take the plant another six weeks to catch up with the fabrication. None of DB's men would be laid off, but would be reassigned to other projects until the steel was ready to roll again. Meanwhile, the cantilevered Span 5 sat gingerly on its falsework with Traveller No. 1 directly over it at Panel Point 4. For once, the bridge was largely silent.

While the bridge was sitting idle, the government awarded Contract No. 3 to John Laing and Son (Canada) to complete the approach roads, overpasses, administration buildings and the concrete roadbed, all for $2,177,267. The cost of the bridge was now over $20 million. Wallace Haughan, president of Laing and Son, expected that two hundred men would be hired within a week and be on the job a week after that: "We shall start the approach roads and overpasses first. Then construction of the toll booths, administration buildings and the final bridge road will keep us busy right up to completion date."[52]

Early on the morning of Thursday, May 19, three weeks before work was to resume on the bridge, Gary Poirier woke in a cold sweat. He had had a nightmare, a premonition that the bridge was going to fall. "In my dream I was looking down on it," he recalled. "I was talking to the crane operator . . . We were both looking at it, and I said, *How come I'm talking to you?* and he said, *I'm dead too.*[53] I didn't want to tell my mother about this dream, in which I was cut in half, because the year before she had had a vision that there were to be three deaths in the family. Sure enough there were; my aunt, my uncle, and my older brother, Ian, who fell off a fire escape and died a week before his twenty-first birthday."

Gary dismissed his nightmare but never completely forgot about it. It was, after all, just a dream, and as dreams do they reflect an individual's own realities; perhaps he had been absorbing too much beer-table talk. Some of the men were worried, and the falsework was the seat of their concern. The legs looked too spindly to support the incredible weight on them, and then there was the wobble, but these thoughts were irrational—the engineers had assured them that everything was okay.

On the morning of Friday, June 13, only a few days after worked resumed, rain pelted down on the North Shore. When the men woke that morning, the forecast had called for mostly cloudy skies, but by the time they arrived at the jobsite the overcast sky had disintegrated into rain showers against the North Shore mountains. Don Heron recalled that "Back east, they used to have this rain vote. The foreman would go out and take a piece of chalk and make a circle on the iron, and if three drops of rain fell in it, you go home, but out here, everybody's got their raingear on and you get to work and then they say, *It's too wet, that's it, let's dog 'er for rain.* We'd spend the rest of the day in the pub."

On that typical wet-coast day, it was either the Lynnie in North Vancouver or the Astoria on East Hastings Street in Vancouver. Both watering holes were familiar, too familiar at times, but on this Friday, Gary Poirier had a hot date that evening and as the Astoria was closer to his home than the Lynnie, they all trooped down there for a few beers and a game of cards. It would be one of their last get-togethers as a full group. The following Monday would also be their last full day on the bridge, but they didn't know that yet, and they still had the weekend . . . life was sweet on the weekend.

6

Collapse

You remember the story about New York when those buildings collapsed there; you called all those firemen and all those policemen big heroes, and they were. But to my heart the guys like Norm, and Bill, and Jim English, and Art Pilon, and Mel Alexander there, they are the same kind of hero.[54]

–LOU LESSARD

The land to the east and west of the north bridge landing is industrial, much of it home to various boat-building and repair businesses, welding and fabrication shops, chemical plants and small manufacturers. At the end of a circuitous route through the western side of this industrial maze, and lying right beneath the bridge deck, is the Lynnwood Marina where a fleet of pleasure boats rock gently with the tide. Surveying this sea of masts is the trendy Marina Grill Restaurant, most of the customers of which are probably unaware of the horrific tragedy that occurred here late in the afternoon of June 17, 1958.

At 2 a.m. on that day, however, Walter Skorodynsky, the caretaker at the marina, was very aware that something was wrong. As he was doing some painting maintenance, he suddenly heard three loud cracks. He initially thought that they were a series of rifle shots fired in quick succession, but on reflection knew they weren't. The sound of snapping steel was unmistakable. He had heard it all too often before while working for MacMillan and Bloedel as a maintenance-of-way foreman on the railroad, and then in the same capacity for the CPR. Glancing up at the dark mass of steel, barely visible in the predawn, he could see that no one was on the bridge, so he turned back to his chores. After thirteen hours of work, he found his bed at 6:30 that morning.

Inspecting the crushed upper grillage of the Bent N4 falsework. PHOTO BILL HADLEY

"I'll tell someone about it when I wake later in the afternoon," he thought, as he closed his eyes.

Dawn broke at 5:06, heralding another warm and sticky day. Although the temperature was to climb to a balmy 25.6 degrees Celsius by late afternoon, a 26-kilometre-per-hour wind from the west would provide a modicum of cooling relief. A couple hours later, when Bill Stroud was pulling his 1957 blue Cadillac Coupe deVille into the curb at the corner of East Hastings Street and Lakewood Drive to pick up Joe Chrusch, the temperature was already 19.4 degrees and climbing. Stroud was single, lived in trendy Kitsilano, and liked the feel of a big car. His friends called it a boat, but they were secretly impressed. Most of the men had families and a Cadillac was only a dream to them. Chrusch was his last stop; he had already picked up Colin Glendinning and Gary Poirier. Chrusch opened the heavy back passenger door, quickly looked in and mouthed sleepy greetings to his buddies.

"Mornin' guys," he said as he slid in beside Poirier, who had his window down, hoping for a bit of fresh air.

"Mornin," Poirier and Glendinning grunted. Stroud nodded and grinned from the driver's seat. All of them were friends, three of them in the raising gang. Poirier, the junior, was the punk, and he took a bit of ribbing for it. Chrusch sat back and relaxed for the ten-minute ride to the jobsite; he had never lived this close to his work before and if he stood on his roof he could probably see the bridge. Working on the bridge was a lot better than one of his previous jobs where he was away from home while building Ben Ginter's brewery at Prince George. Chrusch felt a bit uneasy about the day, but kept his own council.

Waking early, he had entered his children's bedroom to say goodbye, which he did every morning. Maureen, his oldest, was awake, but the other three Chrusch children—Diane, Gordon and Linda—were still asleep. Joe leaned down to give her a quick kiss and Maureen told him that she loved him as he quietly slipped from the room. Leaving the house, Joe got partway down Lakewood Drive before turning back. He entered the house and whispered to his wife, Margaret, "I don't want to go to work today. But I have to . . . I don't know why, but I just don't like this day."

The Chrusch family (left to right): Margaret Chrusch, Linda, 5; Gordon, 6; Maureen, 10; and Diane, 9; with their Christmas puppy, Bobo. COURTESY CHRUSCH FAMILY

"Oh, this is silly," he finally said, laughing, and started back out the door. He didn't get far before returning. Re-entering the house, he gave Maureen a big hug. "Remember sweetheart, I will always love you," he said as he left the house and strode quickly down the block.

Stroud eased his car into the traffic and headed east up East Hastings, made a left at Cassiar and motored north toward the inlet before the road swung east alongside the concrete storage bins of the Alberta Wheat Pool, the most direct route to the old Second Narrows Bridge. Driving around the rutted road north of the newly minted Pier 17, dust boiling from the big whitewalls, he waved to a couple of guys he knew as he turned the corner to the old bridge. The lift span of the old girl was down, which was a relief: clear sailing. As the car moved out over the east side of the bridge deck, the tires made a soporific whirring noise over the steel grate, marking their progress. The men glanced to the west at the new bridge. The tide was ebbing, draining Indian Arm to the east, and they could see the rips and boils eddying around the bridge piers as the water swept out toward the mouth of Burrard Inlet. They were proud of their achievement: the steel was advancing smoothly, but they couldn't take their eyes off the false-work. Was it safe? Poirier's nightmare three weeks previously was still vaguely fresh in his mind, but he didn't say anything. Neither did Chrusch, whose pre-monition that morning was far fresher, and perhaps more alarming. He had a wife and four children.

As Stroud pulled his car into the parking lot on the north bridge approach, the four men quickly exited the vehicle. Usually they went to the Lynnie for cof-fee before work, but on this day they came right to the jobsite. Of the seventy-nine men working on the bridge that day, including twenty-five Boshard painters, many carpooled, so there weren't that many cars or pickup trucks in the tempo-rary lot. Then, hardly any families owned two cars, so carpooling was common. Boshard's men were easily identifiable: white coveralls smeared with red lead or silver paint. Nobody could mistake them for ironworkers, but these men were also not your run-of-the-mill commercial painters. They were bridge painters who were used to heights. They also had the respect of the ironworkers, who con-sidered them one of their own.

Many of the ironworkers were already stripped to singlets or had no shirts on at all, showcasing muscles developed from years of heavy lifting. There were few gyms in those days and being muscular said volumes about who you were and what you did for a living. Other than the stifling heat, it was a perfect day for working outside, the kind of day that the men would drag from their memories on a cold and wet day in January. Monday was out of the way and they were now into the swing of the workweek. Most of the men were already dreaming about having a swim later, or a cold beer on the deck when they got home.

Work was progressing nicely. On June 11, two days after work resumed, Traveller No. 1 had been winched forward from Panel Point 4 where it had been sitting idle for six weeks on top of the falsework, Bent N4, to Panel Point 5. Only five days after that, on Monday, June 16, it was advanced another two panels to Panel Point 7. During the morning and early afternoon of Tuesday, June 17, the men had already erected one bottom chord, one top chord, one diagonal and two verticals of Panel Point 8. Span 5 was very close to being tipped down onto the next set of falsework, Bent N5, which they hoped to reach before noon the next day. Everybody was anxious to land the span.

At lunch that day, Lou Lessard sat with Jim English, Norm Atkinson, Colin Glendinning, Joe Chrusch and Gary Poirier in addition to the dozens of other men that were onsite. Everybody took lunch at the same time and the talk usually centred on the job before dissolving into some good-natured ribbing. There was absolutely no hint of what was to come in a few hours and most of the men were upbeat that the span would soon be safe on another set of falsework. It would be a relief, especially to Poirier and Chrusch.

After lunch, Poirier resumed his frenetic pace, feeding the connectors with whatever they needed. He was also assisting the bolt inspector, Fred Leenstra, to check the torque on the bolts. He generally worked from a wooden catwalk built on top of the lower lateral bracing. With its plank-deck floor and two-by-four railings, it made access to the front end safe and easy, especially for those who were not that adept at walking the iron—like the many photographers and reporters who wanted a risk-free view of what the raising gang was up to.

It was Lou Lessard's job to lead the media out onto the catwalk to show them how the bridge was being built and what the men were doing. He was often bemused by the anxious expressions on some of their faces as they watched the agile connectors lean out into the void to grasp a tugger line, ride the ball or hammer in a drift pin while balancing precariously on their wooden-plank floats. Most of the media were in love with the romance of it—big tools, the growl of heavy equipment, danger, awesome setting—while the men, now inured to it, were too practical to even remotely consider their jobs romantic. Ralph Bower, a rookie *Vancouver Sun* photographer, remembers his tour of the bridge and the one-page article that the *Sun* ran with his photos. It was a big project that had the rapt attention of the city.

Rolf Johnson was the carpenter responsible for maintaining the catwalk. He was with the Carpenters and Joiners Union 452 and had been on the bridge about a month, having replaced another member who had applied for a different position on the bridge. Rolf was an easygoing guy who loved his job, the best part of it being that people left him alone. Just below the northern end of the bridge was a small shack where Johnson kept his tools. It was also where he made up the various components of the catwalk that kept pace with the steel.

Above the catwalk, on the bridge deck, Murray McDonald and John

Gary Poirier, apprentice ironworker, signalling at the front end. Before the advent of hand-held radios, the men used hand signals to communicate with the operating engineer running the traveller. PHOTO BILL CUNNINGHAM, *THE PROVINCE*

McKibbin were keeping close tabs on the erection progress. McDonald was frequently at the front end, not only because it was his responsibility to manage the project, but simply because he loved to watch the steel being erected. He had been heavily involved with steel erection since he was a student engineer on the Lions Gate Bridge twenty years earlier, and liked to be where things were happening. McKibbin, meanwhile, was responsible for taking the many elevation readings required to ensure that the piles and falsework were not settling under the mounting weight building over them. He could usually be found heading down under the bridge with his tripod balanced on his shoulder. Art Pilon, the engineer's assistant, was usually with him.

Art, who had been working as a surveyor's assistant for Swan, Wooster & Partners before signing on with DB, had been immediately assigned to work with McDonald and McKibbin as a rod man. One of his primary duties was to assist McKibbin with his elevation readings. Pilon was delighted that, at the age of thirty-two, he had now found solid employment. He had come a long way from the dairy farms of southern Quebec where only fifteen years earlier he was making the princely sum of twenty-five cents a day in the biting cold of winter and fifty cents a day in the mosquito-swarm of summer. Now he was making significantly more than that an hour. He liked working with McKibbin and thought the job interesting.

In order to take accurate elevation readings, rod tapes had been glued to Pier 14 and to both legs of the falsework. Rod tapes were lengths of paper marked in tenths and hundredths of feet that McKibbin would sight on with his level to determine if the falsework had settled in relation to the pier. The falsework had only settled a minuscule 6.4 millimetres, so McKibbin and Pilon were not expecting anything different today.

A little after 3 p.m., the three men, who had been in the mobile office on the north bridge approach, started to walk out to the front end. McKibbin and Pilon were about to go below to take their readings while McDonald was headed out to see Jim English for a chat. McDonald, McKibbin and Pilon reached Pier 14 where McDonald stopped to talk to Bill Moore, an erection foreman, who was working with Sam Rouegg, a welder, on the jacking assembly on top of the pier. As Span 5 would soon be lowered onto the next set of falsework, Bent N5, the jacks had to be prepared. McDonald reached down to touch one of the tie-down bars as they chatted.

"There's quite a bit of uplift on the links," Moore said.

"Yes, there is," McDonald replied.

"About how much do you think?" Moore asked.

"About two hundred tons," McDonald said.

"That's a lot of weight . . . do you think it's excessive?" Moore asked, somewhat for Rouegg's benefit.

"No, definitely not . . . they can withstand up to nine hundred tons," McDonald replied.[55]

Rouegg couldn't hear the conversation clearly over the noise of his welder, but similar to Chrusch, he too had had a premonition that the bridge was going to fall. So insistent was he, in fact, that he had rattled Moore about it that morning. Rouegg had even asked McDonald point blank.

"Is the bridge safe?"

"Yes it is," McDonald replied confidently.[56]

Although somewhat relieved, Rouegg still had a nagging doubt: *Why had Murray just talked to Bill about safety? Something about the falsework?*

Rouegg turned back to his work while just to the north, Gordy MacLean was carefully engaging the locie's forward gear, advancing the small locomotive slowly along the track until he heard the satisfying clunks of the two bogies budge ahead. As the little train moved out onto the steel with its 52.8-tonne payload, the lower west-side chord, MacLean sounded the whistle to warn of his approach. Meanwhile, Lessard, Atkinson, Glendinning, Stroud and Chrusch, the raising gang, were busy moving their rigging to the west side of the bridge where the chord was to be erected. Quitting time was just over an hour away, at 4:30, and if they hurried, they could just make the connection. If not, however, they would stay behind to finish the job. The chord was the last significant piece before the span would land on Bent N5. In the morning, they would erect two bottom laterals before beginning the jacking process to land the span.

A few minutes later, MacLean set the brake on the locie, a signal to Lou Lessard to instruct his crew to begin preparing the chord to be hoisted. Jack Thompson and Bill Wright, assisted by Alan Stewart, the apprentice, hopped up onto the bogies and dragged the two long chokers of 5.08-centimetre cable already hanging in the hooks of the traveller's lifting block, toward the lifting plate attached to the centre of the chord. While they were doing that, the raising gang was hanging tugger lines, arranging air hoses for their impact wrenches, and seeing to various other bits of hardware required for the connection. They were all in a hurry; it was a brilliant afternoon and nobody wanted to be working much past quitting time.

McDonald left Moore and Rouegg and moved toward the front end to speak to English, who was standing beside the most northerly of the two bogies watching the hooking-up gang prepare the chord to be hoisted. Behind him was the traveller. McDonald nodded to English and said, "Well, what do you think?"

"Well, everything's going along pretty nicely," English replied, casually. "I can't remember a job that I've felt more comfortable with."[57]

"What about the falsework?" McDonald asked.

"Well, the only concern I have," English answered hesitantly, "the only doubt I ever had—and I'm not concerned because you designed it so it's good enough for me—but the only thing that I don't like are the timbers in the grillage."

"Goddamn it, that's exactly what High said," McDonald replied as he banged his fist down on the handrail. "High said 'The beams are alright—I'm just surprised there's timbers in there instead of iron.'"

McKibbin and Pilon were standing nearby, watching the hooking-up gang. McKibbin turned toward Pilon and said, "Well, we'd better get going and take the readings before quitting time." The two headed back along the bridge deck toward the stairs. Below them was Traveller No. 3, which only a few minutes earlier had hoisted the lower bottom chord from the railcar that had delivered it directly from the plant, up onto the bogies.

As they were walking north, Byron Maine, DB's paint inspector, together with Juergen Wulf, Boshard's painting foreman, were just stepping up onto Span 4 after inspecting the painter's work below. They had been making sure that the paint bubbles were being scraped right down to bare metal before paint was reapplied. Earlier in the day, Allan Kay, chief engineer of G.S. Eldridge—an inspection and testing firm that had been contracted by the ministry during the construction—had pulled his inspection crew off the bridge because the Boshard painters weren't prepping the steel correctly. Maine and Wulf had stopped for a moment to watch Anthony Romaniuk, who was hanging in a bosun's chair on the east side of the bridge while blasting the steel with an air gun to remove scale preparatory to painting. They had only planned to be at the front end a few minutes, and had agreed to meet back at the paint shack on the north bridge approach to sign the men's time cards before they left for the day.

As McKibbin and Pilon stepped down onto the ground, they nodded to the carpenter's assistant who was at the entrance to the shack. McKibbin turned and said to Pilon, "Don't bother coming out. I'll just run out and take the readings and then bring the instruments back." As Pilon watched McKibbin walk purposely down toward the shore and the trestle that would take him out to the falsework, he turned to talk to the assistant.

They were quickly joined by Rolf who, at about 3:30, had fished around in his carpenter's pouch for some nails and discovered that he had only about six left—not enough to do much with. He thought, "Oh well, I haven't had my break yet, I'll take it now." Leaving the front end, he quickly made his way back along the catwalk, up the ladder to the bridge deck and north toward the staircase that dropped to the ground near the carpenter's shack. As he was grabbing his thermos bottle, Walter Skorodynsky, the marina caretaker who had just returned from the store after having his breakfast, was about to get ready for his afternoon shift.

Up on the bridge deck, English glanced quickly at his new Gruen watch, a gift from his wife Ruby on their tenth anniversary, and turned back to the activity at the front end while McDonald made his way back toward the ladder that would take him to the lower level. English assumed that he was going down to have another quick look at the falsework. English was concerned that McDonald appeared worried. Bill Lasko, the safety boatman, had already taken McDonald and McKibbin out to the falsework in his 7.0-metre steel rescue barge just a couple of hours earlier. Although nothing untoward was discovered, English sensed that there was an inordinate amount of interest in the falsework that day.

Lou Lessard, who had been helping the raising gang, also looked at his watch and turned to Atkinson, saying, "It's late, I don't think we'll have enough time—it's pretty near 25 to 4:00. By the time we pick it up and boom it down it'll be quitting time."[58]

"Yeah," Atkinson agreed, though Lessard made no move to halt the process.

Frank Hicklenton got the signal from the hooking-up gang, who were still preparing the chord for the hoist, that it would be a few more minutes before the piece was ready. He communicated the delay to Geisser in the traveller's shed, causing him to flick the levers toward him to dog the boom and the main load, as well as tramp the pedal to dog the bull wheel. Stepping out of the shed for a breather, he looked up at the boom, which he noticed had been difficult to swing to the east all morning. He thought that likely the guys hadn't shimmed it quite level. He had told McDonald and English about it earlier in the day and McDonald had decided to stay after shutdown to plumb the bridge to see if it was level. He was probably going to ask McKibbin to assist him.

McDonald would never get that chance, however, for what happened next was beyond his or anybody's knowledge, control or imagination. The tragedy that unfolded—and that had begun, innocently enough, as two dimensional errors generated a year earlier in the little mobile engineering office on the north bridge approach—was now about to become Vancouver's greatest industrial accident.

At precisely 3:40, Walter Skorodynsky heard a roaring noise and rushed to the door of his house at the marina. He shouted to his wife, "They shouldn't allow

No one could have believed that two spans of the partially completed Second Narrows Bridge would collapse with such tragic consequences. COURTESY LOU LESSARD

these jets to fly so low!" Pushing open the door, he stared in disbelief at the bridge as Span 5 began to fall.

The noise the men heard on the bridge was entirely different. Some described it as gunfire or an explosion, others a rumble and still others as a loud click or snapping sound. Whatever their description, all agreed that it was indescribably loud—though Phil Gamble, an operating engineer shaking out steel on the north bridge approach, heard nothing over the engine of his rubber-tired truck crane. But Chuck Robinson heard it loudly from his sister's lawn a few kilometres away, and people all over Vancouver stopped what they were doing to listen: *What was that noise?* they wondered. Jim Bisset, boat builder, was just walking down to his dock a couple hundred metres west of the bridge when he heard a loud cymbal-like clang. Looking up, he saw the bridge fall, stop briefly and then continue its downward plunge. In his mind, it was like a slow motion film; it took an age to reach the water.

With no warning, the south end of the cantilevered span being erected, Span 5, dropped approximately a metre—though the claimed length of that travel would differ from man to man—paused momentarily, and then continued its downward crash to the inlet, carrying men and equipment with it in a maelstrom of twisted steel. The momentum of the collapsing span dragged the columns of Pier 14 two metres to the south, cracking the base of the pier and causing the south end of the 86-metre deck truss span, Span 4, to break away from the north end of Span 5 to which it had been tethered with tie plates. Almost as soon as the bolts on the tie plates sheared with explosive reports, Span 4 lost its grip on Pier 14 and followed Span 5 to the floor of the inlet, all of this occurring within the seemingly impossible confines of six frightening seconds.

Norm Atkinson, at the front end, had no time to think, but his reaction was instinctive. Grabbing the bracing running between the top chords beneath his feet, he rode the steel into the water. Atkinson has little recollection between standing on the bridge deck talking to Lessard and hitting the muddy bottom of the inlet sixty-seven metres below, but he quickly discovered why he had a quick release buckle on his tool belt. Cutting his belt, his life jacket carried him to the surface. Fortunately, Atkinson was paying attention a few days earlier when Ken Bainton, a Fleck Limited and Safety Supply sales rep, donned a life jacket and a full ironworker's tool belt and leapt off the trestle into the inlet. His life jacket held him afloat. Atkinson was now thankful for that, but still a little shell-shocked. He thought, *I'm not going to make it.*[59]

As he was thinking that, his buddy Bill Stroud popped up right beside him, minus the front soles of his boots, which had been knocked off upon impact. Stroud's only thought as he was falling was about the money in his pocket, the dregs of his last pay: "All of a sudden I was in the water, and then the first thing I thought about: *Where's my wallet?*[60] It was absolutely pitch black . . . You couldn't tell up from down."[61]

Both injured men grabbed the same tugger line dangling from the wrecked steel, the flooding tide twirling them around like flotsam in the stiff five-knot current. It was hot and as Atkinson recalls, being in the water was not that bad. Atkinson looked at Stroud: his face was black and blue. Stroud looked at Atkinson and made the same observation. Bodies were floating by them, bright-yellow life jackets propping up comatose and deceased men as the tide hurried them east. Ominously, there were also damaged life jackets with no occupants. Stroud turned to Atkinson and asked, "Do you think that those guys are going to be saved?"

"I don't know," Atkinson replied. "They're going too fast."[62]

Lou Lessard had a floating sensation when Span 5 dropped out from beneath his feet. The span was accelerating far faster than he was, which left him dangling in the void for the millisecond it took for gravity to find him. He had been standing on a piece of steel two metres above the deck between the chord and the traveller where he had a commanding view, after deciding that they would proceed with the connection after all despite it being late, and was ready to give Hicklenton the signals that Geisser would need to guide the piece home. He was watching the top of the boom to make sure that it was elevated enough for the lift and had his head tilted back when the span lurched.

The span dipped on its heavy side first, rotating slightly west under the weight of the traveller, the locie and its bogies, and the 52.8-tonne chord. Lessard immediately lost his balance, dropping directly into the water. Although he had no tool belt on, the momentum of his freefall carried him to the bottom of the inlet. Blood flooded his mouth and he felt the sharp agony of broken limbs as he hung suspended at the bottom, his life jacket having burst off him upon impact. The crash of the span had stirred up so much mud at the bottom that all was black. Lessard couldn't tell which way was up, but after taking a draught of salt water, he finally caught a glimpse of light penetrating the darkness. Pumping toward it with the one arm and one leg that weren't broken, he finally broached the surface. Coughing, gasping and spitting up a briny mixture of seawater and blood, he grabbed a nearby plank rocking in the now turbulent water, the impact of the span having created a mini tsunami that rippled east and west along the surface of the inlet. Clinging to it with his good arm, it offered a temporary refuge and time to think. *What had gone wrong?* he winced, looking around him at the chaos.

Colin Glendinning, who had been working right beside Atkinson and Stroud, fell backward through the air, the pinging sound of snapping cables accompanying the roar of the dropping span as he windmilled toward the water. A cable whipped against the side of his head, almost severing an ear. As he was going down, he thought, *My God, I wish I had a parachute.* He hit the water with his head and shoulders, the straps of his life jacket bursting apart. Hugging the tattered remains to his chest, he rose to the surface, his leg shattered, his chest bruised. He could barely breathe.

No one knew exactly how the last member of the raising gang, Joe Chrusch, fell, but suffice to say that he didn't survive the plunge, fulfilling the painful destiny that he had envisioned for himself earlier that day. His daughter Maureen

Pier 14 tilted, Tower-of-Pisa-like, two metres to the south. OTTO LANDAUER OF LEONARD FRANK PHOTOS, JEWISH MUSEUM ARCHIVES LF-35855

believed that he was fated to die in water, perhaps because of his earlier experience falling through rotten lake ice.

While the raising gang had been on top moving their gear to the west side in preparation of the connection, Gary Poirier had been down below having just crawled out of another chord where he had been checking the torque on some bolts when he was startled by the shriek of the locie's whistle. Nearby, Gordy Schmidt, another apprentice, laughed and said, "If noise gets your goat up here, it's goodbye Charlie."[63]

Poirier was now walking north along the catwalk ahead of Fred Leenstra, planning to spend the last hour of the day topside while Leenstra went to the office to make notes on the bolts they had just inspected. They hadn't gone far when the span took its initial drop. Poirier thought in a detached sort of manner, *Hey, this bridge is going down.*

There was no time to be afraid or contemplate his predicament, but when the span continued its fall, he feared the worst, thinking, *God Gary, you're . . . you're . . . you're dead.*[64]

After that he blacked out, coming to about six metres under water, his life jacket barely clinging to his battered body. Poirier's years of swimming at the YMCA had conditioned him to the water, so he slowly pulled toward the light above him, grabbing a two-by-four plank at the surface. All around him things were still falling from the bridge in a deadly hail of loose bolts, kegs of bolts, oxyacetylene tanks, welding machines, planks, rails and tools. Each one of them could prove fatal.

Wow, you're not dead, he thought. *Your life jacket is almost ripped off—you gotta' stay afloat. You're hurt.* The tide whisked him and his plank east, eventually fetching him up against the piles of the old Second Narrows Bridge 120 metres away.

"Come up here," a voice shouted down at him from the bridge deck.

"I can't get up," Poirier replied, exhausted and in pain.[65]

As Poirier was looking up at the men hanging over the rails of the old Second Narrows Bridge trying to dangle a rope his way, Charlie Geisser was nursing his wounds at the end of Span 5. Although he had grabbed the lines coming off the drum to stabilize himself when he first felt the bridge jar, he had been tumbled about on the steel during its decent and was now sitting hunched over near the drowned end of the span. He had narrowly missed being crushed by the chord when the traveller's cables had tightened on the massive piece of steel, which had then dropped as the bridge dipped and the bogies it was sitting on fell away from beneath it into the inlet. One of his

The mangled traveller at the end of the collapsed Span 5. OTTO LANDAUER OF LEONARD FRANK PHOTOS, JEWISH MUSEUM ARCHIVES LF-35510

colleagues, Frank Hicklenton, lay dead nearby, crushed beneath the traveller's bull wheel.

Jim English was alive as well. Fortunately his life jacket had held when he hit the water feet first and shot right to the bottom. Rising to the surface, he grabbed a floating plank, and although he was hurt from both the fall and a flying plank that had smashed into his face, he was not that badly injured. He recalled, "All the way down I kept saying, *It can't happen, these bridges don't fall down!* We had such faith in the engineering, that it [seemed] impossible."[66] English would later joke that he learned to swim that day.

Others were even more fortunate. Byron Maine and Juergen Wulf were in the middle of Span 4 when they heard a sharp report and then a rumbling sound. Maine initially thought that it was the locie moving back and yelled at Wulf,

"Come on, let's get off the tracks, the locomotive is coming back." Almost immediately, however, they realized it was not the locie. Although Wulf was six-foot-six, Maine, who was much shorter, and pudgy, matched Wulf's strides step for step as they raced for the safety of Span 3, which they reached just as Span 4 was going down. Once safe, they looked back at the disaster. Their view, which was initially obscured by a voluminous cloud of dust boiling off the top of Pier 14, was quickly restored as the prevailing westerly began to shift it east. The miasma quickly dissolved into chaos.

Bill Moore and Sam Rouegg, who were perched on top of the columns of Pier 14, rode it like a bucking bull as it vibrated and then was pulled over by the dropping Span 5. At the first sign of vibration, Moore leapt to protect himself against a concrete parapet at the top of the columns and was quickly followed by Rouegg who grabbed onto Moore's waist to prevent himself from being catapulted over the edge as the top jerked two metres south. This very spot was the one that Rouegg, because of his premonition, had picked out earlier in the day as being the most protected site, but Moore had beat him to it. Moore recalled that "There were rivets snapping off and steel grinding against steel. We looked for a place to hide. There was no place to go. Some of the overhead steelwork crashed

Sam Rouegg's and Bill Moore's perch on top of the tipped Pier 14. PHOTO
P. STANNARD, DOMINION BRIDGE CO. LTD.

down on top of the pier and toppled into the water. Scaffolding fell around us and then No. 4 pier [should be No. 14 pier] started to go the other way. I didn't see very much but I could hear what was going on. Then everything was still. I've heard about 'deafening silences.' Now I know what they're like."[67] The silence, however, quickly surrendered to a horror of anguished cries coming from below.

The steel was still shifting, making a tortured grinding and screeching sound, but that was almost background noise to the unforgettable wail of human agony. Lessard recalled that men were "crying, screaming, crying for help, either dying or afraid of dying."[68] Gordy MacLean had gone over with the locie and bogies, and amazingly had survived the plunge. He was now immersed in water up to his neck, with the tide rising quickly and his leg jammed inside the crushed cab. He was desperate. A short distance away, big John Olynyk was poking his head up through a diagonal and yelling for help: "Get a cutting torch, or I'm done for!"

Although not badly injured, he was trapped; the diagonal's manhole was at the sunken end of the member. Fortunately he had found an access hole that was just large enough for his head, and it was from there that he was urgently requesting help. The tide was already at his waist and rising. It was a miracle that Olynyk was alive at all. Only two minutes earlier a member of the bolting-up crew who had been in the diagonal, Alan MacPherson, had yelled at him to switch places. It was a fortunate switch for Olynyk but not so for MacPherson, who perished. As the bridge was crashing down, Olynyk thought for sure that he was gone. All he could think about was that he would never see his wife, Edna, again.

A bit of dust in the eye saved another member of the bolting-up crew, Ted Barkhouse. Barkhouse, who had been bolting up the upper west-side chord, excused himself to go to the first-aid shack on the north bridge approach to get some dust particles removed from his eye. He was on his way back to the front end when he felt the vibration and saw the front-end traveller move south and then dip below the horizon. Span 4 quickly followed, but not before Barkhouse had found his legs and safe haven in the middle of Span 3.

His partner, Dennis Gladstone, was just as fortunate. At about 3:30 he had asked Rod Smith, the bolting-up gang foreman, for permission to leave the bridge to go downtown on some personal business. His request was granted and he was driving south over the old Second Narrows Bridge just opposite Span 5 when he heard a couple of loud reports like a rifle being fired. Turning quickly, he was shocked to see the bridge collapse. Earlier that morning, when he was on the east side of the front end bolting up a diagonal from the inside, he had felt the bridge take an unusual dip and then recover. He was used to the bridge swaying and bouncing with the actions of both the locie and the traveller, but this movement

"Twenty-four hours and we would have landed the span," Colonel Swan lamented on the day of the collapse. OTTO LANDAUER OF LEONARD FRANK PHOTOS, JEWISH MUSEUM ARCHIVES LF-35477

was so unusual that he remarked about it to Barkhouse: "Did you feel that?" he asked, astonished.

"Yeah, I did," Barkhouse replied, equally mystified.

Don Gardiner, another member of the bolting-up gang, felt the bridge lurch and knew instinctively that it was about to fall. He leapt over the edge, plunging deep into the inlet. Releasing his tool belt he shot to the surface where he met a falling plank, which smashed into his shoulder, dislocating it.

Anthony Romaniuk, who was working on the same side of the bridge where Gladstone felt the dip, thought like Byron Maine that the vibration was caused by the locie moving back. But when it got a little more serious, he knew that the bridge was in trouble. Riding the steel down into the inlet, he found himself still dangling from his bosun's chair, although now he was well under water.

Yanking unsuccessfully on the rope to free himself, he then realized that he was still harnessed to the chair. Releasing the harness, he quickly made his way to the surface.

Murray McDonald was presumably back on the bridge deck when the span dropped, arrested its fall for a fraction of a second, then fell again. Colin Glendinning remembered him climbing up the ladder from the lower level just minutes before the collapse. One can only imagine McDonald's shock in the last few seconds of his life. McKibbin was also in a precarious position, likely directly beneath the bridge, when it collapsed. Both men were killed instantly, a somewhat morbid blessing given that neither could have borne the shock of surviving. Mary McDonald, on first hearing the news, remarked to their daughter Diane, "Let's pray that he's dead, because he couldn't live through this."

McKibbin was probably the only man within close proximity to the false-work when the webs of the upper grillage beams suddenly folded north, the upper flanges crushing right down to the level of the lower flanges. The process likely started on the west side, which was more heavily loaded. The failure was accentuated by the plywood "softeners" above and below the stringers, which had begun to creep north. Immediately after the webs of the upper grillage on the west side failed, the load simultaneously shifted to the east leg, where the webs of that side of the upper grillage were also crushed flat. The span was now in motion, creating a dynamic load that exceeded the bearing capacity of the legs, which quickly folded north approximately seven metres above the level of the lower base plate.

In a sad twist, McKibbin had already packed up his instruments and was probably only minutes from moving out from under the bridge. Pilon recalls that "he must have had the instruments maybe already to go because the tripod, the two tripods, the legs were inside one another." Pilon would find McKibbin's tripod early the next morning at the bottom of the inlet near where he had been working. That same day, McKibbin was certified as a fully registered BC engineer, with "very high recommendations by the board of examiners."[69]

McKibbin's mother, Nita Wilkinson, woke in a restless state at her home in Australia, on the Wednesday night, having just had a dream that something had happened to their son. "She dreamed he put his arms about her and looked very troubled,"[70] Reverend Sam McKibbin, John's father, would later recount. John's wife Barbara had just finished teaching at Queen Mary Elementary School on Trimble Street in Vancouver where she was preparing her students for a concert the following day. When she got on the bus at Fourth and Trutch to go home, the driver informed her that the bridge had fallen. In her heart, she knew immediately

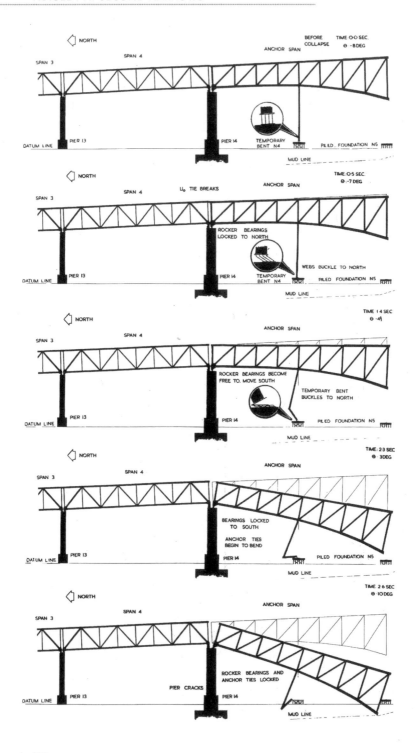

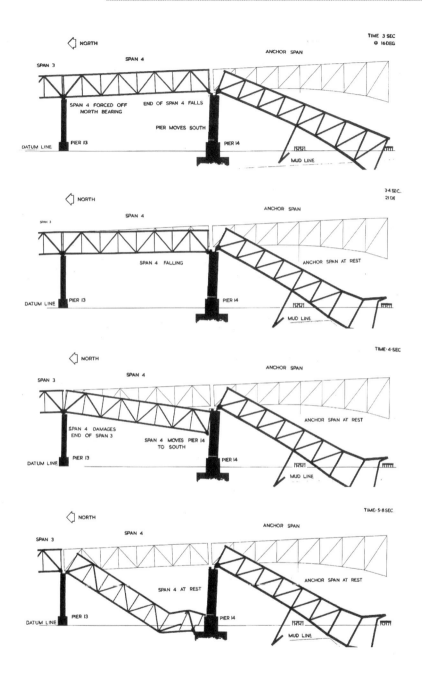

The collapse occurred within the seemingly impossible confines of six frightening seconds. BURNABY PUBLIC LIBRARY

that John was gone and that he, like McDonald, would have been unable to deal with the tragedy had he survived.

The Reverend Sam McKibbin echoed those sentiments when he later wrote to Barbara: "I feel that had he lived after the crash he could have been crippled for life. Even if he knew he was not to blame I do not believe that he would have ever recovered."[71] In a magnanimous and selfless gesture that brought a lump to the throats of all who read it, the reverend also extended his sympathy to the people of Vancouver: "But we are thinking not only of our boy and his sweet wife. The remembrance of the many stricken homes in your city through this calamity fills us with compassion for those who have equally suffered, and in many cases more deeply suffered."[72]

The news telegraphed quickly around the city. Ironworkers' wives heard about it from other colleagues' wives, friends and neighbours or, as Mary Mc-Donald did, over the radio while she was doing the laundry. She didn't know at that time whether her husband was alive or dead. It would take a police officer at the door to confirm the latter. Margaret Chrusch could not discover the fate of her husband despite trying, and even wives who arrived at the bridge site were denied access. Some wives rushed frantically from hospital to hospital only to discover later that their husbands would not be coming home, that they were already at the Vancouver City Morgue.

The news arrived at DB's various offices perhaps a little more swiftly. John Prescott was sitting with High Carpenter in his office at the plant when the call came through that the bridge had fallen. Total disbelief was soon replaced with the mind-numbing reality that men had been lost, spouses had been widowed and the company was embarrassed. Angus McLachlan was devastated—McDonald had been his friend. Tudy Grelish, contract supervisor, remembered that the daughter of one of the women in the office phoned to tell them that the bridge had collapsed. Tudy wondered, angrily, at the time, "Why didn't someone from DB contact us rather than let us hear it from the street?"

The sirens made it impossible to concentrate. *Who was on the bridge that we know?* all the office staff thought. Hugh Dobbie, domiciled in the Hastings Street drawing office, remembered the sirens as well, but just before that, his project leader had received a call from DB's purchasing agent: "The bridge is down!"

"But as this guy was a bit of a comedian," Dobbie recalled, "we all said, *Oh yeah, right*, because we're all working on the bridge, but then we heard the sirens racing up Hastings Street and we all got in our cars and headed over to the bridge."

Downtown, in the old Standard Building on West Hastings Street, a young

Span 5 collapsed over the N4 and N5 falseworks. COURTESY PEGGY STEWART

Vancouver lawyer, Tom Berger, was sitting at his desk reviewing some files when John Graham, one of the firm's lawyers, summoned him to the window. "Does it look like the bridge has collapsed?" he asked, astonished. Berger looked incredulously at the dropped spans, never even imagining that a year later he would be defending these very ironworkers against an injunction granted to DB during a legal strike.

Just down the street at the Terminal City Club, Glen McDonald, the city coroner, had just finished a late lunch of Dover sole and Stilton cheese, and was still basking in the company of good friends when news of the collapse shocked the dining room. Everyone rushed to the roof to catch a glimpse of the bridge. McDonald's first concern, after adjusting to the horror of it all, and the inevitable work, was a jurisdictional one, so he phoned his equal in North Vancouver directly from the club.

His colleague, who was not one to waste time on formalities, made the decision easy for him: "We can't handle a case like this. We're not equipped for it. If it is a bridge collapse, it's an absolute bloody disaster! There'll be so many bodies we couldn't possibly do the work here."[73] McDonald hurried back to his office to face what would surely be one of the most difficult days of his career. The deceased had to be identified, families notified, autopsies performed and an inquest organized, the latter with the consent of the Attorney General at Victoria.

Hundreds of similarly astonished faces were pressed against hundreds of other harbour-view windows. One of those belonged to John McKibbin's good friend, Graeme Kelleher, an engineer who had graduated with McKibbin and ventured with him to Canada to seek the experience of a new land. Upon seeing the bridge down, he immediately phoned DB and was told that McKibbin had been on the bridge at the time of the collapse. Kelleher instinctively knew that the chances of survival were slim, but he still wasn't prepared for the shock of losing his best friend. Kelleher and McKibbin were well aware of the risks of bridge building; they had co-authored a joint thesis on the Sydney Harbour Bridge where sixteen men were killed during its construction. That knowledge, however, did little to lessen his pain.

As fast as the news hit the street, it attracted the attention of the press. Radio announcers, TV personalities and the print media were quickly all over the story. But in the *Vancouver Sun* newsroom, it was a lazy afternoon. It was hot and the evening edition had already gone to press. Frank Rasky, editor of *Liberty Magazine*, wrote about the scene:

> Certainly, nobody in the newsroom of the *Sun* expected the biggest scoop of the year to break. The deadline period was passed then, and a few of the late afternoon shift reporters were standing around, waiting to go home.
>
> The phone rang. Because the receptionists had all gone home, Jim Hazelwood, a reporter who covers the waterfront, answered it. After agreeing, "Oh yes," a couple of times, he put the receiver down.
>
> "A woman says the Second Narrows Bridge fell down," Hazelwood said, with a laugh. Nobody got excited.
>
> "We thought the woman was either joking or mistaken," recalls Audrey Down, a crack *Sun* reporter. "My first thought was that some joker was trying to play a trick on us. Like the joker a year before. He phoned up a local radio station to say the Second Narrows Bridge was on fire. After publishing an excited bulletin, the station had to correct

it. Even after the calls started flooding in, and pandemonium began to mount around this red-hot story, we still thought it was the old Second Narrows Bridge that had done the splits.[74]

Edward Herring heard the crack of steel while driving over the old bridge in a gravel truck: "I thought they'd let off some dynamite. But when I looked I saw about twelve men leaping from the falling spans. Then the bridgework seemed to vanish in a cloud of smoke. I stopped my truck and I could hear the men in the water hollering for help as the tide swept them under the old bridge toward Indian Arm."[75]

Ralph Bower had been covering an event at the Pacific National Exhibition when he checked in with his office and was told to get over to the new Second Narrows Bridge, pronto. Like his newsroom colleagues, he initially thought that it might have been something to do with the old bridge, perhaps a ship collision, but as he raced over that bridge he could see that it was something far more serious.

Activity around the construction site was frenetic. Burrard Inlet was beginning to jam with tugs and pleasure boats and when Ralph pulled into the parking lot, ambulances were arriving in droves. As he got out of his car, he met his boss, who said breathlessly, "I'm out of film."

"So I gave him three quarters of my stock." His boss then directed him to take his Speed Graphic to the waterfront where men were being dragged from the water, some dead. "You go down to the waterfront," he said, "and I'll cover the rest."

Jean Howarth, *Province* reporter, was crossing the old Second Narrows Bridge a few minutes after the collapse when she looked down and saw the flash of a bright-yellow life jacket moving east. She wrote that "The riptide swept the first dead man into view as we were running across the Second Narrows Bridge. The tangle of the new bridge was there, but it hadn't seemed real until then. He seemed real. His hands were caught in the webbing of his life jacket, and he floated easily, gently, out behind the trapped hands, with his head under water. He wore very big, heavy boots. They seemed to brake him in the moving tide . . . Sight of boots was what hurt."[76]

Ed Cosgrove, a reporter/photographer at *The Columbian,* was also on the old bridge when he saw the two spans collapse:

> I heard the rumble and then almost piled head-on into the car ahead of me . . . I got a flashing glimpse of the driver's face as he twisted in his

seat and pointed. I followed the direction of his point and saw the two outermost spans of the partially completed bridge lurch slowly forward, then plunge into the fast-running water . . . I reached the shoreline at a dead run, just behind the driver of the other car . . . We saw three men floating in the water. Two of them clung feebly to pieces of wreckage floating in the water.[77]

Cosgrove also recalled the experience in an interview for Glen McDonald's autobiography:

I was the only cameraman there. The guys were struggling to get out of the water but most of them couldn't move, they were pinned or injured too badly. They were yelling, screaming. It was gory as hell. People were pulling them out . . . I kept shooting until I ran out of film. I've always felt guilty for not trying to help the guys. I shoved the used roll in my left pants pocket and reloaded. I was running down the pier when I felt the roll slip down my leg. I watched it fall into the water and sink. Christ, what a feeling! I'd forgotten there was a hole in the pocket I'd been after my wife for weeks to mend. I've thought since that maybe it was God's way of punishing me for shooting instead of helping.[78]

One of Cosgrove's bosses, *Columbian* co-manager Wen Ballantyne, was simultaneously in Victoria on a press junket when the news was passed to Gaglardi by the press. "The conference was about to open when a white-faced Gaglardi said he had just got word of the disaster which took the lives of sixteen bridge workers."[79]

As Ed angled for the best shot, crowds were beginning to gather behind him, staring with morbid curiosity at the carnage while police officers endeavoured to keep them at bay. Hurrying along the shore to capture the rescue as it unfolded, he stopped to record a sad reminder of the finality of it all, a priest kneeling to administer last rites to one of the men.

7

Rescue

*The collapse of the bridge is nothing. Steel can be replaced,
but we can't replace the men who lost their lives today.
Naturally the government is going to find out the causes but
the important thing right now is to alleviate the suffering of the
injured and the families of those killed.*[80]

—PHIL GAGLARDI

At 3:40 p.m. the Canadian Coast Guard Cutter *Mallard*, a 12.2-metre ex-RCAF steel crash boat, was slowly cruising by Burnaby Rock at the entrance to Vancouver Harbour, having just slipped under the Lions Gate Bridge on its regular tour of Burrard Inlet. The wheelhouse radio was tuned to CKWX where the popular DJ Red Robinson had just received a call from his uncle, Chuck, informing him of the bridge collapse. Don Sinclair, a Canadian Air Force para-rescue specialist, was on board, having just been assigned a marine shift while stationed at Canadian Forces Base Comox. Almost simultaneous with Red's announcement, marine radios all over the harbour began to squawk with the news. Everyone on board had an *Oh my God* expression glued to their faces, looks that soon turned serious as the crew quickly prepared their equipment and themselves for what would surely be a complex rescue and recovery mission.

As the captain of the *Mallard* pushed home the throttles of the two big 165-bhp diesel engines, lifting the surging craft at the bow, men at the bridge site, who were just recovering from the shock of the collapse, were now galvanized into action. Art Pilon, who was just coming to the realization that McKibbin had inadvertently saved his life by insisting that he stay at the carpenter's shack, raced to the water, foolishly discarding his life jacket as he ran, forgetting that

Men scramble over the twisted steel to rescue their injured comrades. *THE PROVINCE*

he couldn't swim. There he found the old night watchman's boat, a little dinghy with a small outboard on it, tied to a small wharf attached to the wooden trestle alongside the piers. The engine started on the first pull and Pilon roared out into the inlet. The first man he saw was Jim English, sitting on the star bracing of the downed steel, his face slick with blood, his arm hanging limply by his side. He had been knocked unconscious upon hitting the water and came to while rising to the surface where he had grabbed a plank and made his way over to the steel. English shouted at Pilon as the little boat turned toward him.

"Pick up the other guys first. I'm fine!" he yelled.

"I need you for weight," Pilon screamed back over the noise of the engine, as he pulled abreast of English. English struggled into the boat and Pilon manoeuvred the craft into the current and through the tangle of steel. English immediately spied Charlie Geisser at the end of the collapsed Span 5, and yelled at him to see if he was okay.

"I'm fine. Leave me alone, guys," Geisser grimaced in response. "I have to get my nerves back straight. Just leave me, I'm fine, just leave me alone . . . Get the other fellas out."[81]

Next, they found Lou Lessard clinging to some planks underneath the span. With English leaning over one side, Pilon pulled Lessard into the boat on the other as gently as he could. Lessard grimaced in pain, blood dripping from his twisted leg, his arm broken. Pilon then ran the boat back to the wharf where he dropped off Lessard and English.

While men were assisting them out of the boat, another saw that Pilon had no life jacket on and insisted that he wear one. Behind him, the downed spans were still groaning and screeching, settling into the mud at the bottom of the in-let. Men were screaming for help, rescuers yelling at them to hang on. "Help will be there soon!" they were shouting. It was an emergency that had no precedent, that could never have been rehearsed. DB had a safetyman and a boatman, but this was far bigger than either of them could ever have imagined or planned for. The man who had been in charge of the bridge, Murray McDonald, lay dead be-neath the wreckage. Lessard and English, now safely on the wharf, provided some direction as they waited for medical help.

Art retraced his path into the twisted steel and found a gasping Colin Glendinning clutching his broken life jacket to his chest, blood dribbling from his mouth and streaming from the side of his head where his ear was hanging, almost severed. His leg was also broken. Back to the wharf with Glendinning and then into the steel again, Pilon found Norm Atkinson and Bill Stroud still deter-minedly clutching the tugger line that was now stretched taut to the east with the strength of the flooding tide. As Pilon was alone in the boat, he quickly devised a plan to get them both aboard without capsizing it.

"Listen guys," he said. "You gotta listen to me or we'll all drown . . . as soon as I come near, each of you get on either side of the boat and then I'll bring you aboard one at a time."

"Okay!" they both cried in unison as they released their grips on the rope and grabbed for the boat's gunwale, Stroud crabbing around to the starboard side. Pilon hauled Stroud in first and then reached over to do the same with Atkinson, but Atkinson's zipper pull on his life jacket was hanging up on the side of the boat. Turning to Stroud, Pilon said, "Lean way over, so I can get Norm in." Pilon then grabbed Atkinson's leg, tipping him into the boat before turning back to the wharf.

Don Sorte, rescue diver, catching a lift on a boat to avoid being swept away by the tide. COURTESY HISTORICAL DIVING SOCIETY CANADA

There, pacing anxiously back and forth was Ernie Duggan, a DB employee who had rushed over from the plant as soon as he had heard the news. He was desperately seeking news of his eighteen-year-old son Kevin, an apprentice on the bridge who had followed his dad and two older brothers into the trade. "I'm looking for my son," he said hopefully to Dennis Gladstone. "Have you seen him? He was an awfully good swimmer."[82] As the distraught father was questioning other survivors, the body of his young son was being brought ashore.

While Pilon was delivering Atkinson and Stroud to the wharf, his wife, who had only just heard the news on the radio, was panicking at her home on West Sixth Avenue in Vancouver. She was imagining the worst. A neighbour rushed her to the old Second Narrows Bridge, but they wouldn't let her across, heightening her alarm, and the Lions Gate Bridge was jammed with the traffic that normally used the old bridge. Desperate, she raced from hospital to hospital, eventually making it to the North Shore General Hospital (NSGH), but still no sign of her husband.

Jim English's wife, Ruby, heard the news from an ironworker's wife whose husband was on another job at the time.

"The bridge has just collapsed," she said excitedly over the phone.

"You're kidding," Ruby replied, first thinking that it was a sick joke and then not wanting to believe it.

Ruby quickly phoned DB and was told, "Yes, the bridge has collapsed, but Jim was seen walking around." In a matter of minutes, her pal Monica Gartley, wife of Stan Gartley, appeared on her doorstep. Monica had already made the rounds of the Vancouver hospitals with Fran Glendinning, but now she and Ruby raced over the Lions Gate Bridge to the NSGH to see if Stan was there.

As the *Mallard* pulled onto the scene, small boats were beginning to gather, their skippers having the best of intentions, but their numbers and the wash from their boats hampering the search. Phil Gamble had run to the water's edge as soon as he could make it down off the bridge approach and got aboard a small private vessel. "There were boats—I can't believe it to this day, where they all came from—tugboats, and so quick, you know," he remembered. East of the bridge, a tug was moving methodically back and forth, fishing men from the water, a number of which lay under a tarp on the stern—a *Vancouver Sun* report called the tug's cargo "Grisly sheaves of death."[83]

The bridge tenders on the old Second Narrows Bridge had a commanding view of the inlet and had already watched a few of the bright-yellow life jackets and their lifeless occupants swirling beneath the lift span like so much flotsam. They helped to coordinate some of the rescue by marine radio, communicating

Tugs and small craft race to the rescue. OTTO LANDAUER OF LEONARD FRANK PHOTOS,
JEWISH MUSEUM ARCHIVES LF-35472

with two tugs, *Marjorie L* and *Davis Straits*, both owned by the L&K Lumber
Company. Ron Sundstrom, skipper of the *Davis Straits*, picked up five men as
far away as 2.4 kilometres east of the bridge, all in their life jackets and all dead.
Harry Chapman, skipper of the *Marjorie L*, recovered one such fatality.

Don jumped off the stern of the *Mallard* onto the steel with a cutting torch
to clear a path through the tangle of cables obstructing the rescue, and noticed
Bill Lasko, DB's boatman, moving around the steel in his bright-yellow rescue
barge as he searched for survivors. It had been Lasko's father, also Bill, who had
fulfilled the same function prior to the shutdown, but during the layoff he had
decided to return to commercial fishing, leaving his son with the safety job. Bill
Jr., who was twenty-five years old, had been idling his boat east of the bridge when
it collapsed. He was now shaking with the shock of losing so many of his friends
and with the responsibility for having to recover them. Adding to his distress, an
errant life jacket had been sucked into the prop of one of the outboards. As Lasko

was peering over the head of the now-silent motor at the shredded floatation device, shocked that it might have held one of his friends, Pilon and Lessard were yelling at him to "Forget about it. Use one engine. It'll be powerful enough." For Lasko, however, there was no objectifying the deaths, no rationalizing that these were people you didn't know or care about. At such a young age, to lose so many, was beyond contemplation.

Nearby Mel Alexander, soaked to the skin, was cutting other parts of the steel to free the pinned bodies of some of his colleagues. Right after the collapse he had begun to assemble cutting torches and gear, and together with his crew, was providing some semblance of organization to the rescue and recovery.

Below him, Pilon was still cruising around in his little boat after having dropped off Atkinson and Stroud. He soon spied Gordy Schmidt sitting in a chord, his head poking out through a hole that was too small to extricate himself, but large enough for him to be eminently visible. Pilon tied the boat to the steel

A *Vancouver Sun* report referred to the tug's horrific cargo as "Grisly sheaves of death." THE *VANCOUVER SUN*

and walked up the chord. Schmidt was sitting calmly, waiting for rescue. His leg, above his knee, was a bloody stump, fifteen centimetres of grisly bone poking from the meat of his thigh. Schmidt had been in the chord when the bridge fell, and when it finally came to rest, had looked down to see that his leg had been amputated where the chord had folded over it. With steely resolve he had removed his belt and cinched it tightly around his thigh to prevent an almost certain fatal haemorrhage. It was only a matter of circumstance that Schmidt had been on the bridge at all. He, Al Snider and Don Gardiner were all roommates, Snider and Schmidt being apprentices. Snider got the call from the Local that day to report to the bridge, but as he wasn't home, Schmidt took his place.

Pilon gave him a cigarette and checked his tourniquet as they waited for the burning torch to come and free him. "Now when they burn this, you step way back," he said. "When the plate drops, wait till we get it cooled off a bit." When the steel was finally cut through, Pilon lifted the 86.4-kilogram apprentice from the hole and carried him to a tug, which had come alongside the steel. Schmidt was not only missing a leg, but he was now burned on his side and his shoulder by the torches wielded to free him. "Got him in my arms like this," Pilon remembers. "I walked down the chord which was about this steep—there was a big tugboat there. I walked right out to the tug, sat down, kept him in my arms. We went around, way around the front, the pier, way down, over the shore, and I walked out of the goddamned tug and up the walkway to the ambulance with him. Now, where the strength came from I don't know."

Nearby, a drama was being played out with John Olynyk, who had been working inside a diagonal when the bridge fell. Tossed around inside the steel on its descent, he suddenly found himself two metres under water. Turning awkwardly around, he began to crawl up the tight confines of the steel tomb when he realized that the entry port was behind him below the waterline. He continued moving up to a small port that was just large enough to poke his head through and it was from there that he was desperately seeking help. The tide was first at his waist, then quickly at his chest and rising. Within minutes, it would be over his head. Two blocks away, six welders—Jim Fullager, Ernie Enderson, Norm Miles, Alex Kellay, John Neilson and Cornelius Walstra—were working for the North Shore Welding and Iron Company when they heard the collapse, and knew that their services would most likely be required. Quickly assembling a mobile cutting unit, they jumped into their shop truck and raced toward the bridge.

Neilson remembered that "There was nothing but confusion when we got there . . . Guys were running around with blood streaming down their faces and nobody seemed to know what to do."[84] Then they heard John Olynyk and were ferried out to where he was trapped. "We worked like hell to save his life," said Fullager. "He just stared at us without a word."[85] As the men cut the steel away, others passed hard hats of cold water to cool it. "Be patient," they cautioned. "The steel's hot, so wait a minute once they cut the plate out . . . wait till it cools." Olynyk didn't wait a fraction of a second after the steel plate dropped from the hole, leaping out of the diagonal like a man possessed, a man who had just discovered he would live after all.

Sitting in the boat, waiting for Olynyk to be freed, was Charlie Geisser, who had finally got his head around the collapse and had made motions that he wanted to be helped off the steel. Dazed, he walked down to the edge of the collapsed span to wait. A rescuer shouted at him: "Stay there, Charlie. We'll come and get you—don't move." When he was eventually escorted to the boat, he suddenly doubled over, clutching his chest in pain: "Oh—my chest, my chest."[86]

Up on the bridge, Juergen Wulf looked down at the activity below, not quite believing that he was safe and that what he was seeing was real. Then he saw one

The rescue barge tied up beside the crushed traveller. COURTESY CRAIG ELLIOT

147

of his colleagues struggling in the water. Running down the span, he found a rope and grabbing it, shinnied down to within ten metres of the water where his burning hands forced him to release his grip. Dropping the rest of the way to the water, his hands bloody from the friction, he was able to assist the man to shore.

Frank Rasky of *Liberty Magazine* recounted the experience of another painter:

> Another heroic painter, thirty-two-year old Laci Szokol, had run out of paint, and was just getting a refill when he felt a *snap, click, snap, click* under his feet. "I run thirty feet along the catwalk just like crazy," Szokol said. However, he stopped to throw down a rope, and pull up his floundering pals, Michael Josefi and Tony Wohlfart.[87]

Wohlfart had just come out of a lower west-side chord, where he had been painting, when he felt a heavy vibration, which he at first thought was the locomotive, but when he turned and caught sight of Span 5 dropping, and then the columns of Pier 14 beginning to shift south, he knew that Span 4 wouldn't be far behind it. He jumped into the inlet before Span 4 fell out from beneath his feet.

Byron Maine, meanwhile, was headed in the opposite direction to get some equipment down to aid the rescue, after hearing an ironworker from below shout, "Get some oxyacetylene and torches down here quick." Together with George "Fergie" Ferguson, the compressor operator and equipment maintenance man, they quickly assembled what would be of most value. As they were doing that, an ironworker joined them.

"How're we going to get this stuff down there?" he asked. "Wait a minute, I see a pickup over there; we'll use that." Hotwiring the truck, they loaded it up and peeled out of the employee parking lot, smoke boiling from the squealing tires as they burned out onto the road that would take them by the Lynnie, along Columbia Street, around a few industrial buildings and under the bridge. The streets were now beginning to choke with onlookers streaming toward the disaster, but the driver just laid on the horn forcing them off the road, vowing to knock anybody down who got in the way.

In another pickup truck was George Hoffman, who together with a couple of other men had recognized the need for more cutting equipment and had raided DB's equipment shack on Main Street for bottles of oxyacetylene and a couple of cutting torches. A carpenter's apprentice, he had been working on the Main Street overpass building concrete forms when he looked up at the sound of the collapse. *Gee, it's odd that on a nice clear day like this we would have thunder,* he thought. Then he noticed the huge cloud of dust drifting away from Pier 14. Looking over, he noticed that the traveller had disappeared: "There was this total silence, country wide for about fifteen or twenty minutes . . . and then the wailing started. It sounded like all of North Vancouver was coming to the rescue—and it was fire trucks and ambulances coming in from all directions and the wailing went on and on and on and on. It went through my mind for weeks."

In one of those ambulances, a converted Ford delivery van, were Charles Todd and his partner, both with the North Vancouver Fire Department (NVFD). They were on Capilano Road in North Vancouver, having just picked up a heart-attack patient, when Bill Thompson, their captain, radioed them.

"Where are you?" he asked urgently.

"We're on Cap Road with a coronary. We'll be back in a few minutes," Charles replied.

"Well, drop him off quick as you can and get over to the Second Narrows Bridge. It's just collapsed," Thompson ordered.

In record time, they made it to the NSGH, admitted their patient, and were barrelling along Keith Road, sirens and lights full on, when they crested the hill and had their first glimpse of the bridge.

"Oh my God," they both mouthed.

They were among the first ambulances on the scene, and within minutes were on their way back to the hospital on their first of many runs. Before they could return, however, other ambulances and emergency vehicles began to arrive from all directions, the unending wail of their sirens splitting the air with a cacophony of urgency. Executives at Hooker Chemical, just east of the bridge, heard the collapse and sent their ambulance and first-aid team to the site. Leaving their ambulance to idle behind them, they and other ambulance attendants rushed to the water's edge and along the trestle with their stretchers to meet the men as they were being ferried ashore. Richard Halloway, another member of the NVFD, was there with an inhalator aboard the department's 1942 pumper truck to try to revive some of the comatose men. He had been off-shift at his home on St. David's Avenue in North Vancouver when he heard the collapse and called in to see if he could help. Madly pedalling his bicycle seven blocks up to the hall, he jumped in the department's truck and drove it quickly down to the bridge site.

Nearby were the consulting engineers, Colonel Swan and his partner, Hiram "Hi" Wooster. Swan's grey shirt matched the ashen pallor of his face as he and Wooster watched stretcher after stretcher being carried up the beach toward the waiting ambulances and ominously black, unmarked panel trucks which were used to convey the dead to the morgue. They were lamenting the fact that within

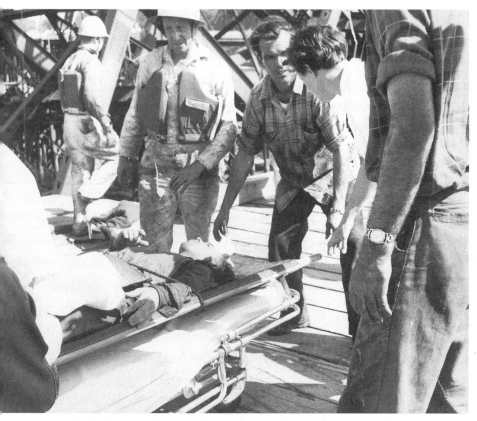

Ambulance attendants on the wooden trestle, attend to an injured man. COURTESY CRAIG ELLIOT

twenty-four hours, Span 5 would have been landed on the next set of falsework and the collapse would have been avoided. When asked about the rumours that in the last couple of days some ironworkers had complained of the bridge being noticeably wobbly at the front end, Swan's comment said volumes: "It was the business of the contractors [Dominion Bridge] to take care of the stresses."[88]

Other onlookers were showing the kind of selfless compassion that emergencies bring out. People were abandoning their cars and running down to the beach to drag the injured men to safety. Ed Olsson, a maintenance foreman on the old Second Narrows Bridge, rushed into the water and together with two other employees, Elmer Wylie and Albert Trewit, rescued three men as they drifted close to shore. They could only watch as other men, beyond their reach, were carried quickly east with the current. On the beach, one woman took off her blouse to cover a shivering victim, another sat in ankle-deep mud in her best clothes to comfort an injured worker. A waitress, dressed in the form-fitting uniform of the day, helped the divers into their wetsuits, later bringing cases of pop from her restaurant to refresh them and the other rescuers. East of the collapse, two doctors who had been out fishing saw Gary Poirier struggling against the pier of the old Second Narrows Bridge and raced toward him.

"I think my leg's broken," Poirier said as they eased alongside him and gently pulled him over the side. After making him comfortable, they asked if he would mind if they motored down the inlet in search of others.

"No," he mouthed painfully, closing his eyes and slowly shaking his head.

George Hoffman witnessed Poirier's partner, Fred Leenstra, coming out of the water: "A couple of guys that I knew fairly well; one of them was a bolt inspector, checking bolts and stuff like that. Young fella, and he said, *Whoooff, what a ride!*"

"He was still clutching the torque wrench," remembered Poirier.

Leenstra, who sustained serious chest injuries and almost bit his tongue off on impact, recalled that "I was going to go to the office walking up the lower level of the bridge. Suddenly I saw everything buckle. I knew it was coming down. I jumped into the water. Then I heard a crash, like an atom bomb or something. A steel lateral bracing—about two or three tons of it—crashed about six inches from me."[89]

The rescue was now beginning to jell. Skip Pratt, the construction administrator, after immediately calling DB with the news, had grabbed a first-aid kit, and together with Charlie Moore, the old sergeant major, had run down to the trestle to provide the first level of care to the injured. Boats conveying the deceased to the north shore were now being redirected to New Brighton Beach on the south shore due to overcrowding. There, the *Mallard* and other rescue vessels unloaded their gruesome cargo, which was quickly spirited away by another fleet of black panel trucks. The sight of the covered bodies being loaded into them caused many to catch their breath—it was the finality of it all—but as the immediate priority was getting the injured men to the hospital, that was what finally brought a measure of order to the chaos.

Andrew Roberts, a police officer with the Vancouver Police Department (VPD), and a member of the department's marine squad, was on holiday at the time, taking a course to upgrade his marine skills, when the whole class was informed of the collapse. Running from the classroom near Victoria Square in downtown Vancouver, he made his way to Coal Harbour where the *VPD 45*, the department's boat, was moored. It had already left for the bridge, so Andrew grabbed his gear from his locker and hitched a ride on a speedboat, which raced him to the site. There, he transferred to the police boat, but Bill Dobie, the captain, wouldn't let the men dive. There were too many boats milling about and there were already several divers in the water. It was too crowded and dangerous, so they backed off.

One of the divers in the water was only sixteen years old. Phil Nuytten had taken a break from school to follow his dream of opening Vancouver's first dive shop. Just as SCUBA diving was becoming popular, he opened Vancouver Divers Supply on Fourth Avenue. It was later relocated to Victoria Drive, where he and two partners made wetsuits, sold diving equipment and did the odd salvage job. To aid the latter, they had an illegal tugboat radio in the shop so that they could scope out any possible salvage jobs and get the drop on their competitors. At 3:40 p.m., Bill Bamford, Nuytten's partner, was in the shop listening to the marine radio when he rushed outside and called to Nuytten, "Phil, come quick, you gotta listen to this." Nuytten, who was already partially dressed to go out on a small job, rushed into the shop just as the radio was crackling and squawking with, "It's falling, the whole thing's falling . . . my God, it's coming down and people are falling," the voice shouted.

Nuytten told Bamford to quickly phone the VPD to inform them, ask for a police escort, and advise them to call the Vancouver Fire Department (VFD), which had a rescue diving team. Jumping into the truck, they raced down Victoria Drive to Kingsway Avenue where a police escort was waiting with lights flashing and sirens blaring. "So we crossed the old Second Narrows Bridge and it was absolutely incredible to see this bridge [the new bridge], the end of it in the water and everything else, so we went on the north side, around behind the Lynnwood Hotel and got in where the falsework was . . . we drove the truck down to the end and I started getting the rest of my gear on and went down to the dock . . . there were already a number of bodies on the dock that they had picked up floating."

Within minutes of arrival, Nuytten was in the water, swimming toward the crumpled locie. Gordy MacLean was still alive in the cab, screaming for help as the tide was already at his neck and rising. Nuytten was quickly joined by two of his salvage-diving buddies, Scratch and Don. Together they tried to gain access from the water to free MacLean's leg. Nuytten made three attempts to swim into the crushed machine, even taking off his tank and dragging it behind him, getting to within two metres of MacLean, but still to no avail. He was just coming up for the third time to scope out another tact when he was told to forget it, that MacLean was gone. Looking over, Nuytten could see MacLean's blonde hair washing gently just under the surface, a terrifying moment now turned so suddenly peaceful. In stark contrast to the urgency, Nuytten noted, somewhat surreally, that butterflies were wafting lazily by in the late afternoon heat.

The tide was now at its peak, sucking the divers east and forcing them to cling to the steel to maintain their positions. Swimming was out of the question, even for Nuytten who was a Canadian spearfishing champion. He was towed around by a boat, but when he was close to one of his objectives, an ironworker who couldn't dump his tool belt, he released his hold to reach for the man. As soon as he let go, the tide slammed him up against the wreckage, a long steel splinter slicing between his tanks, piercing his wetsuit and pinning him to the steel. If he had hit it at any other angle he too would have become a casualty. When he finally extricated himself, the man was gone, having been picked up by another boat. Below him he could now see the bottom, the majority of the disturbed mud having been swept east with the current. Two ironworkers were directly beneath him, held against the steel by the pressure of the current. Weighed down by their heavy

Boshard's safety man talking to rescue diver, Phil Nuytten, clinging to the steel.
PHOTO BRIAN KENT, *THE VANCOUVER SUN*

tool belts, they were suspended there, not quite sitting and not quite standing, their arms outstretched as if reaching for help, their hair rippling in the current as though it were the wind.

Overhead, an RCAF helicopter scoured the water, searching for any signs of life, while on the trestle, viewing the horrific scene with Colonel Swan was Highways Minister Phil Gaglardi, who had just arrived by float plane. Finding

no safe spot to land in the inlet due to the crush of small boats, Gaglardi's pilot had diverted to Vancouver Airport where the minister and his entourage rented a car to race him to the site. Even then he was delayed near the bridge when an RCMP officer demanded to see his identification before permitting him entry to the site. Not much more than an hour earlier he had been in a meeting discussing the Queensborough Bridge with New Westminster municipal representatives, when the news of the collapse was conveyed to him by the press.

"You're joking," he responded, in disbelief.

When assured that they weren't, he immediately conferred with Highways ministry engineers before being instructed by the premier to fly to Vancouver. Before he left Victoria's Inner Harbour, he expressed his condolences to the families of the lost men. He also stated that the tragedy would delay the bridge by about six months and that

Minister Gaglardi and Hiram F. Wooster survey the damage. *THE PROVINCE* the collapse would not cost the province "a nickel more."[90] He assumed that DB carried the appropriate insurance.

By the time Gaglardi and Highways officials had made it to the bridge site, the rescue had already become a recovery mission, as all those who had had a voice had been pulled from the wreckage and the water and had been taken to the various hospitals around the city. As Gaglardi viewed the downed spans, he mused that "It was just one of those things. It seems impossible."[91] Colonel Swan, whom Gaglardi had once described as "the best consulting engineer in Canada," would only add that "A commission will investigate the collapse. Until that happens we won't know the answer."[92] Speculation, however, that the falsework was to blame, was already gaining popularity.

Bill Cunningham, a reporter with CBC television news, was on the scene, and in the dramatic style of the day, brought his candid interview with Gaglardi into our living rooms:

> **Cunningham:** Spans 4 and 5 of Vancouver's Second Narrows bridge plunged into the waters of Burrard Inlet this afternoon with a tremendous hiss and a terrible loss of life. Hours after the disaster the death toll still isn't certain, but it is certain that it is one of Vancouver's worst industrial accidents in many years.

> **Cunningham:** What are your reactions after having inspected the bridge?

> **Gaglardi:** Well, it's a pretty tragic situation. As far as the steel is concerned and the cement, that can easily be replaced. That's not worrying me too much. The thing that's worrying me are the families that have been hit by this tragedy and the men that are killed, and I'd like to take this opportunity of sending condolences to them. We certainly sympathize with them and will do everything we possibly can to alleviate all the difficulties as far as we can.

> **Cunningham:** Any idea, sir, about the cause?

> **Gaglardi:** There's a number of different things that have been suggested as to cause, but we have here the best consulting engineers on the west coast, Swan and Wooster, and steel erectors, the Dominion Bridge. I don't think you can get anybody that's a better firm as far as

steel erection is concerned, and it's just very, very difficult to know what the cause is and we're going to have a thorough investigation and we might be able to find out the cause. We're all interested in knowing, but it's pretty hard to say right at this moment.

Cunningham: Any chance of any more of the bridge coming down, sir?

Gaglardi: No, I don't think so. I think it's pretty solid . . . But it will be thoroughly inspected so that there certainly will not be any public hazard of any kind. That's one of the things that we must take very careful inspection of.

Cunningham: How about plans for the future of the bridge, sir?

Gaglardi: Well, the work must go on. We'll tear down what's been destroyed, salvage what we can and continue the job. That's the only thing we can do.

While Gaglardi had been squinting into the glare of the TV camera spotlights, across the Strait of Georgia Premier Bennett had been busy taking immediate strategic action to establish a royal commission to investigate the tragedy. After summoning the Attorney General, Robert Bonner, he announced that Chief Justice Sherwood Lett of the BC Supreme Court would be appointed to head up the investigation.

The leader of the Opposition, CCF leader Robert Strachan, not to miss a political opportunity, demanded a full-scale investigation of the whole public works department. He also questioned "whether the accident could have resulted as a result of laxity or interference by Highways Minister Gaglardi," referring to the ministry's difficulties with keeping personnel:

I think it is now obvious that the number of first-class engineers, which Mr. Gaglardi's policies drove out of government service, created a vacuum which he has been unable to fill . . . It must be obvious that a major change in policy is required to bring back the department . . . some of the technical, trained personnel, who can in future prevent a recurrence of this disaster . . . The Minister of Highways, whose department is responsible for this work, has been under question by the opposition for some years now . . . We have felt that as a layman he interfered much too much with the technical personnel, trained in highway and bridge construction, and out of this interference has come the classic expression that Mr. Gaglardi's ideas are a triumph of imagination over engineering . . . Is this bridge collapse a result of Mr. Gaglardi's imagination?[93]

Gaglardi shot back, "As far as my department is concerned, or my administration, or my engineers, anybody can investigate any time they like because we've got nothing to hide."[94] He went further in a telephone interview with the *Kamloops Sentinel*:

> Let Mr. Strachan put up or shut up! The statement attributed to the CCF leader that as a layman I interfered too much with the technical personnel on the Second Narrows Bridge is a lot of bunk. I say this—if Mr. Strachan can prove his charges, let him fire away. But I'll explain how the whole thing works. In the meantime, it's just a cheap political trick capitalizing on a tragedy where many lives have been lost and right now we don't know how many more . . . To start with, the bridge isn't ours yet. Not until the contractors finish it. Which also means the taxpayer's dollar is fully insured until the job is done. Much will have to be done over again . . . Also, one of the top engineering consultant firms—Swan and Wooster are in charge and Dominion Bridge are doing the building . . . You think I'm going to interfere with those fellows? Right now both those companies feel terrible about what's happened. No one knows what caused the collapse.[95]

Premier Bennett struck out at Strachan as well, calling his comments "irresponsible," and that "He's trying to make political capital out of this tragedy. It's just the usual wild statement from a wild man."[96]

In Ottawa, Public Works Minister Howard Charles Green stated that the tragedy was a "reminder of the dangers always faced in construction work and of the courage of these men who play such a great part in developing our country."[97] Prime Minister John Diefenbaker extended his sympathies to the victim's families, as did the leader of the Opposition, Lester B. Pearson. And in direct contrast to the actions of the leader of the provincial CCF party, CCF MP Harold Winch stated sympathetically that the "tragedy hit him very deeply."[98]

Minister Gaglardi, meanwhile, had more immediate things on his mind.

After visiting the bridge site, he put on his pastor's hat when driven to the NSGH to visit the injured men. There, Ralph Bower was recording the event. Some of the men looked up and were surprised to see the affable minister at their bedsides, grasping their hands and wishing them a speedy recovery. "Yes, Uncle Phil came by to shake my hand," Jim English recalls. "He was very supportive." Another man who came by to shake English's hand was John Heron, Don's dad. John had been on the bridge only a few days before the collapse when he had had an argument with English who promptly fired him. Now he was at English's bedside: "I want to shake your hand. Thanks for firing me," he said.

Across the inlet at VGH, Fran Glendinning, after four hours of searching, had finally located her husband Colin. Hours earlier, she had turned on the TV for her children to watch cartoons after school when news of the collapse had flashed across the screen. Now she was at VGH where the associate director of administration, George Ruddick, had only a few hours earlier rolled out the hospital's recently designed disaster plan. He had only had ten minutes notice after receiving an excited call from his wife telling him that the bridge was down.

Eileen Haynes, in charge of emergency nursing at VGH, was bracing herself for an onslaught of injured men. She remembers walking down the corridor outside the emergency ward where most of the men were lying on gurneys, there being no more room in emergency, when she saw Glendinning with his broken leg and partially severed ear. He looked up at her and mustered enough strength to say, "Hi cutie." No one would let Fran see him: it was too frantic and there were too many emergency cases for hospital staff to be worried about the sentiments of a distressed spouse. Doctors were quickly assessing the worst of the twelve cases, prioritizing them for immediate surgery, while in the hospital's waiting room, ambulance crews and police officers were consoling anxious and grieving relatives. Like many other worried spouses, Fran was given a few sleeping pills and told to go home. "Somebody will call," they said.

At the NSGH, similar scenes were being played out. When informed of the collapse by one of the nurses, Dr. Don Warner sought out his surgeon colleague Dr. Bill Arber, and the two of them leapt into a car and raced down to the site. Dr. Arber hopped aboard a boat to look for survivors while Dr. Warner attended to a man on the beach who was already beyond help. Soon, the two realized that they would be of more use at the hospital where the men were being taken.

There another colleague, Dr. Emol (Pete) Therrien, an obstetrician who had once been an army surgeon and who was now in charge of the hospital's triage unit, was quickly assessing the trauma cases and assigning them to various physicians. He assigned Don to a young French Canadian with a severely fractured

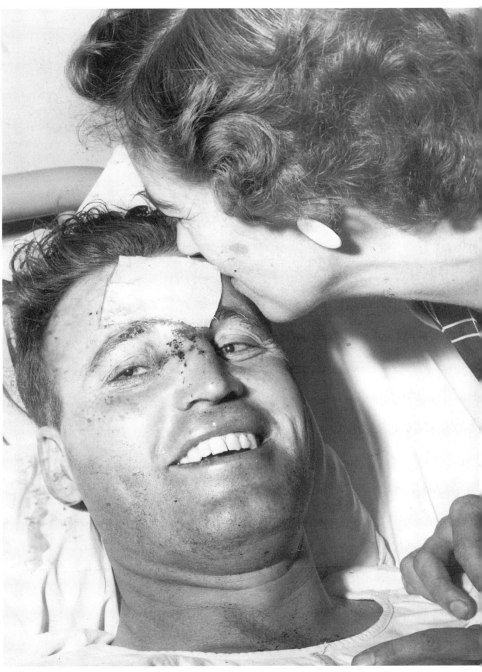

Ruby English kissing her injured husband, Jim. COURTESY JIM ENGLISH

femur, Lou Lessard. "I remember him telling me a horrendous story," Don recalled. "[The span] dropped and he was standing on it, and the bridge dropped faster than he did, so he was floating above the bridge and he ended up in the water and he found some drift that he got on . . . He pulled himself up on it, but his leg wouldn't come so he reached out with his arm and pulled his leg up."

While Lessard and seven of his colleagues were being attended to by NSGH staff, in the waiting room wives, relatives and friends were gathered, offering each other support, waiting for prognoses, trying to make sense of it all. They all knew that ironworking was dangerous, but each had learned to suppress the fact that one day their husbands might not be coming home. Their spouses had cheated death this time, so for them, this was not that day. Other wives, however, were also there, some near collapse. They were being told that their husbands weren't there; they had already been to VGH and had been told the same thing. They had also been turned away from the bridge site. The awful truth was now just beginning to sink in. There was only one more place to look: the Vancouver City Morgue.

Back at the bridge, divers were working furiously to recover bodies before they were swept away by the tide, which had now turned and was ebbing west. One of those divers was Jack Bridge with the VFD, who had been down at Second Beach with his wife and kids, and had just stepped in the front door of his house when the phone rang. Duncan, the dispatcher, was on the other end.

"Jack," he said, "we've had a bad accident. The Second Narrows Bridge has just collapsed."

"Come on Duncan," Bridge said, "surely, you're kidding."

"No, I'm not," he said, seriously. "Rescue and Safety's at the Terminal Docks with your gear, and a boat will be there to take you to the site."

Within minutes of arrival, Bridge and his partner Len Erlendson were in the water, moving toward the downed spans. Almost immediately, Erlendson pointed to a man in a life jacket hung up on some steel below the surface. It was Murray McDonald. Struggling with the current, they got a rope around him and managed to get him to the boat where Simma Holt, a *Vancouver Sun* reporter was on board to record the tragedy. She was horrified. The rest of the day was spent scouring the bottom and the steel for the remaining four men, two of whom were located but not recovered: painter Rudy Holzl, apprentice ironworker Alan Stewart, ironworker Richard Mayo and ironworker Stan Gartley.

Bill Lasko, DB's boatman, was still on site late that afternoon. When implored to take a break, he would only say, "No, I saw a shirt with a life belt still on it, floating out there. It belongs to a man I know. I am going to find him."[99] As

Len Erlendson (left) and Jack Bridge, rescue divers, search for the missing men.
PHOTO BILL HADLEY

the afternoon dissolved to evening, men like Lasko, Pilon, Bridge, Erlendson and dozens of others were forced to quit, thankful that they had been rescuers and not casualties, thankful that they could now go home to their families, unlike the men they were searching for. When Bridge arrived at No. 2 Fire Hall in downtown Vancouver later that evening to hang up his gear, he couldn't help but notice activity at the back of the nearby coroner's office, inside of which the bodies of the men were being neatly laid out pending identification and autopsy.

Art Pilon too, had his moment. When he finally walked into his house at 10 p.m., friends and neighbours were already gathered in his living room, consoling his wife who believed that he had perished. Upon seeing him, she broke down. Pilon, who had rescued so many, was forgiven his transgression for not calling. Even later that same evening, one man was driving steadfastly through the night to Vancouver from Kelowna to discover the fate of his father and to assist with the recovery. Donny Geisser, Charlie's son, was working for DB as an ironworker on the Kelowna floating bridge when he got news of the collapse. DB didn't know the status of his dad, but Don knew that his services as an experienced diver would be required. He rushed home, got his wife and child, and drove all night to Vancouver where he dropped his wife off at her mother's place.

After determining that his dad, who was in hospital, was okay, he drove immediately to the bridge site. As he exited his vehicle, he was confronted with a painful reminder of the tragedy. Cars and trucks of his lost friends were still sitting in the parking lot on the north bridge approach, and in the lunchroom, thermoses and lunch buckets were awaiting owners who would never claim them. He began diving right away, but within a couple of hours the tide began to flood and the current became too strong. He and the other divers got out of the water to wait for high slack tide. Men were still missing and Don and his colleagues were determined to find them. It had been Alan Stewart's dad, Sid, who as shop foreman in 1952 had first introduced Don to the ironworking trade. Now Sid's son was missing and Don wanted to help find him.

Diving would continue for days after the collapse, though by this time the search was confined to the steel when slack tide permitted work on or near the bottom. On June 25, eight days after the disaster, the body of Boshard painter Rudolph Holzl was brought to the surface by Jack Bridge. Rudy's cousin Hans, with whom he had shared a room, had the difficult task of identifying his remains at the Vancouver City Morgue, but this was not as painful as the call he had to make to Rudy's widowed mother in Austria to inform her that her son would not be coming home.

When everyone thought that the bridge couldn't possibly take another life, on Thursday, June 26, it added its last victim to the roster of the dead. Leonard K. Mott, one of the recovery divers, who was what they called a hard-hat diver, had only recently been introduced to the skill of SCUBA diving. One of his previous jobs was as a stand-in for Peter Lorre during filming of Jules Verne's *Twenty Thousand Leagues Under the Sea*.

Mott was working for Seaboard Marine Divers on the Deas Island Tunnel, the same project that had created so much furor in Victoria with Gaglardi and Neil McCallum, when he got a call from the diver's union to report to the Narrows. He was a ticketed diver, and as SCUBA divers didn't have their own union at the time, the ironworkers wanted a union man if there was to be any cutting done. Another diver, Dave Arnold, a professional SCUBA diver who was already working at the site, was competent, but as he wasn't ticketed he was restricted to assisting Mott.

Arnold and Mott were about nine metres beneath the surface, cutting into a chord in search of the missing bodies. As Mott was cutting through the steel, slag was building up on the cut, holding the plate in place. Arnold signalled to Mott

From left to right: Bill Hadley, Jack Bridge, Bill Parks and Len Erlendson prepare to dive on the collapsed spans. PHOTO BILL HADLEY

that he was going to find a hammer to knock the plate out, but when he returned, Mott was gone. Although no one will truly know what happened, it is suspected that Mott ran out of air—he was only wearing a single tank—and that as he made his way to the surface he hit a cable, which dislodged his mask, possibly disorienting him. The fact that he had also taped a chunk of welder's glass over the front of his mask probably didn't help his vision any.

Jack Bridge speculates that he was too heavily weighted, which would be normal for a hard-hat diver, but unusual for a SCUBA diver, and that he didn't have a quick-release buckle on his weight belt. Bill Hadley, another VFD diver, suspected that he had put his tank on over his weight belt, making any quick release—if he had had such a device—almost impossible within survivable limits given that he was most likely already out of air.

Surfacing about thirteen metres east of Bill Lasko's rescue boat, time was against Mott as he was swept toward the old Second Narrows Bridge with the flooding tide. The equipment tenders, who were still anchored over the wreckage, saw him wave and struggle against the current, his mask down over his mouth and nose. They thought that he was waving that he was okay, and watched as he tried to reinsert his mouthpiece. They threw him a line, but it was too late. A few moments later, with a great splash he disappeared just under the west side of the old bridge. The bridge tender on the old structure watched Mott struggle and then suddenly sink from sight. It would be the last that anyone would see of Leonard Mott.

The next day, June 27, Monica Gartley finally received the call that allowed her to sleep: the bridge had finally given up her husband, Stan. He had been lodged in one of the chords where he had been working. Donny Geisser brought him out. "I had to take my tank off and then drag it behind me and grab him and pull him out," he recalled.

In the hospital, the men were now beginning to feel their pain. Norm Atkinson's neck was cracked, every rib was broken and his testicles were the size of grapefruits. He felt as if he had been through a war—but at least during the Battle of Britain he had only sustained a leg wound. "Peace time can be dangerous," he mused. Lou Lessard's leg was in traction and it would be months before he would be well enough to go back to work. But go back to work he would, even if on crutches.

Bill Stroud was hospitalized for eight months with a compressed fracture of his back, several broken ribs and chipped teeth. John Olynyk and Gary Poirier were checked over and released that night, but Poirier returned the following day complaining of a sore leg. He would wear a cast for weeks. Jim English was

Len Mott, once a stand-in for Peter Lorre during filming of the movie, *Twenty Thousand Leagues Under the Sea*, attending to hard-hat diver, Matt Mathews.
HISTORICAL DIVING SOCIETY CANADA

released quickly as well. The bridge needed a superintendent to begin the dismantle, and English was going to be there for it. Nothing essential would be touched at the bridge site, however, until the royal commission had concluded its investigation.

There was no question about most of the men returning to the job. It was in their blood, and besides, High Carpenter had asked them to. He was one man they had a hard time refusing.

"Norm," High said to him at his hospital bed, "you have to go back and dismantle the bridge."

"If we hadn't have gone back then, then we would have never gone back," Atkinson recalls.

Across the inlet, Gordy Schmidt, minus a leg, was clandestinely liberated from the hospital by his roommate Al Snider, and with blood still spotting his stump bandages, was rolled downtown for a few quick ones. Schmidt would never dance the iron again. At VGH, Fran Glendinning finally got to see her husband, his leg encased in plaster, his head bound with bandages, both lungs so badly bruised that he lived on a quarter of one for three weeks until they healed. Exactly one year to the day later, he would break the other leg and be attended to by the same nurse and doctor at the same hospital. Not long after that, Glendinning would leave the trade forever, when he and a few others attempted to form their own Canadian Ironworking local against the wishes of the International.

On July 6, 1958, at Empire Stadium, a memorial service conducted by Reverend George Turpin, D.D., Shaughnessy Hospital chaplain, was held to honour the deceased and to pay tribute to the survivors. "We remember single acts of bravery when someone will respond to a sudden emergency," he said. "But today we are remembering the daily courage of men whose tasks take them into dangerous places and pay tribute to them. They all shared a common danger and now they will live on in our hearts and minds."[100]

As the long roll call of the dead was read, a single red rose was picked up from a table by a colleague and placed gently over a golden ribbon. After the ceremony, a cortège of mourners accompanied the nineteen scarlet roses to the Narrows where they were cast, one by one, near the twisted steel that had claimed so many. As the blossoms swirled away in the tide, crimson against the grey cast of the water, they slowly disappeared from sight, but not from memory.

8

The Coroner's Inquest

*"My God, the bridge has fallen down!" It was so unreal, so
unthinkable, because there was only one bridge that conceivably
could have fallen down and that, of course, was the new second
crossing of the inlet in the harbour.*[101]

–GLEN MCDONALD WITH JOHN KIRKWOOD,
HOW COME I'M DEAD?

Glen McDonald, the City of Vancouver coroner, took his job seriously. He knew
that his inquests were, "the last hearing a deceased is going to have," and instructed
juries with that fact in mind. "Ask of yourselves that one last question: *How come
I'm dead?* The dead could not ask it. The jurors could, on behalf of the deceased.
Maybe they could find the answer."[102]

At his office at 240 East Cordova Street in downtown Vancouver beside the
Vancouver City Police station, late that tragic afternoon, McDonald was mulling over the reasons why he had accepted the responsibility for the dead of the
Second Narrows Bridge collapse. The NSGH morgue was too small for so many,
as were many other hospital morgues, which were essentially designed to accept
those who had died in their care. "Spouses going to a hospital might think that
their loved ones were still alive and being treated," he reasoned. "No, the morgue
was the best place to hold and identify the bodies and conduct the inquest."
Given that this was the worst industrial accident in Vancouver's young history,
McDonald was understandably aware of the importance of his role in quickly
investigating the deaths and determining their cause.

Meanwhile, the deceased were being delivered to the back door of the morgue
in the alley behind the building. The *VPD 45* dropped several bodies off at the

foot of Main Street just three blocks to the north, but many others came by ambulance and black, unmarked panel trucks, some from New Brighton Park and others directly from the North Shore. Even though some of the bodies were unrecognizable, McDonald had to make them as presentable as possible for viewing by the next of kin without destroying the cause of death, which could only truly be revealed later through autopsy by pathologists Dr. Tom Harmon and Dr. Eric Robertson. "Each body was on a stretcher, with clean sheets and pillowslips, made as cosmetically acceptable in appearance as the tragedy allowed,"[103] he later wrote.

Dealing with grieving relatives is always difficult for a coroner, but it would be insensitive to lead a spouse, mother, brother, sister, father, son or daughter to the wrong loved one. Therefore, someone had to identify the bodies prior to a relative and who better than the men's employer, Dominion Bridge. McDonald called DB and Swan-Wooster: "The plan was to get the bosses at the Dominion Bridge Company and Swan Wooster Engineering to send their superintendents and foreman down to the morgue to make at least some preliminary identifications."[104]

Skip Pratt, at twenty-one, was far too young to bear this burden, but as he was the construction administrator on the bridge and knew most of the men, it was, unfortunately, his lot. "I did everything. I looked after the billings and the customers, I looked after the hiring for all the projects, I looked after labour negotiations, I looked after safety, I looked after the construction shed, loaded and unloaded," he recalled. In that list of duties there was no mention that he would ever be responsible for attending the morgue to identify remains, but that is exactly what he was tasked with—even though, in some instances, the bodies defied identification in the normal sense. Although some of the men had already been identified and were wearing the ominous toe tags, Pratt would confirm the identity of those and try to find names for the men whose identities were still a mystery.

McDonald later wrote of the horror that Pratt faced and would never forget:

> The appearance of a body crushed by a girder or a steel beam is not pleasant to look at, not even for medical people used to such sights. We had them all that night. There were multiple fractures; there were compound fractures; there were limbs missing. It was far worse than what we got from automobile accidents. We were dealing here with an average weight of 165 pounds for a body and the crushing force of girders

pinning these poor men. It was like jaws of a huge steel machine with a force that couldn't possibly be resisted. It's awful to think about but some of the men were almost truncated—cut in half, crushed in half. Some were no longer human. They looked like abstract Picasso paintings. Twisted, garish, unreal . . . The memories of that night still haunt me.[105]

"Most of them I recognized," Pratt recalled, "but if some of them didn't have their name tags on I don't think I would ever have recognized them."

Early that evening the coroner's courtroom began to swell with grieving and fearful relatives who had come to the morgue as a terrifying last resort when their

"We were dealing here with an average weight of 165 pounds for a body and the crushing force of girders pinning these poor men. It was like jaws of a huge steel machine with a force that couldn't possibly be resisted." Glen McDonald, Coroner. COURTESY CRAIG ELLIOT

spouses couldn't be located at any of the local hospitals. Consoling them was a host of ministers, whom McDonald had invited to comfort the bereaved as well as to give last rites to the departed. He had immediately called St. James Anglican Church across the street and asked them to put their coffee pot on and make some sandwiches. He then called the Salvation Army, the United Church and the Catholic Church, hoping that he had covered all possible denominations.

As each anxious spouse was escorted into the viewing room, nurses from the jail next door were on hand and morgue attendants hovered closely with ammonia capsules gripped tightly in their palms to revive them in case they collapsed from shock. It was over quickly, and soon after 11 p.m. the morgue returned to its normal sombre state. The coroner's staff, however, was not free to go; they still had a full night of autopsies ahead of them. No stone could be left unturned. Toxicology tests had to be performed: were the men influenced by sleeping pills, pep pills, alcohol or carbon monoxide? This information, although seemingly ridiculous in the face of the obvious disaster, would be required by insurance companies and perhaps would become evidence at the lawsuits that seemed likely at the time.

Early the next morning, the fifteen autopsies complete (four men had yet to be recovered), McDonald sat down to organize the inquest, which he planned to open on June 24. McDonald was understandably concerned about the makeup of the jury. Not only would it have to be impartial, but the jurors would have to have a technical background in order to understand the inevitable bridge jargon that they would hear. Usually, juries are composed of randomly selected members of the public chosen from tax roles, but in this case, McDonald reasoned, he would need a jury that was already familiar with bridge and engineering terms.

Where else to find such a group of special adjudicators than by canvassing local unions for a list of eligible men and then doing the same with local engineering firms for some management types? He would, of course, steer clear of DB and Swan-Wooster, which could not help but be partial. None of the twelve lawyers representing the deceased, the unions, the companies and the provincial government, had any objection to this course of action, and a jury of two engineers and six ironworkers was duly appointed. They were soon to hear that of the fifteen men who had died and been recovered, autopsies confirmed that eleven were killed by multiple injuries while four had drowned. It was important to note that the trauma deaths were found to be mercifully quick.

Although McDonald proclaimed that he would endeavour to "stay as close as I could to the rules of evidence, omitting hearsay," he explained to the jury that "the only hearsay would be the evidence of the various experts who were known to vary in their opinions as to why the bridge collapsed."[106] It would be up to the eight jurors to sift through the facts and sometimes-contradictory opinions to winnow out the truth, thereby determining whether the deaths were due to accident or any other means. "If possible," McDonald instructed them, "you can make recommendations to prevent such an occurrence in the future, and if fault is to be found, you can determine that as well."

McDonald addressed the circumstances of the case in his autobiography, *How Come I'm Dead?*:

> It was a case of the old Latin expression used in law, *ipso loquitor*, which, when translated, means the thing speaks for itself. After all, bridges aren't supposed to fall down when they're being built. It seemed there must have been negligence somewhere so I knew there would be many arguments and disagreements . . . The testimony concerned itself with such questions as the tidal effect in Burrard Inlet, a possible major earthquake, a sliding change of the surface on which the temporary bridge footings were placed, poor design and lack of accurate follow-through in the construction.[107]

Before about two dozen spectators, including two women who had rushed from the courtroom in tears during the medical testimony, Jim English, still sporting two black eyes and facial cuts, was called as a witness. "Everything was normal,"[108] he said simply of the condition of the bridge just before the collapse, noting that most of the men who had died had been working near him. Donald Mitchell, who had only just joined the bolting-up crew the day before the collapse, and who had ridden the falling Span 5 into the inlet without serious injury, reiterated English's statement. The inquest carried on for almost two hours that day before adjourning at 11:40 a.m. to allow the jury to review the evidence.

The verdict they reached was predictable. The deaths were "unnatural and accidental and that another man died of accidental drowning while looking for some of the missing bodies underwater."[109] With respect to the cause of the collapse, they reasoned it couldn't be determined without further investigation by professionals. No recommendations were therefore forthcoming with respect to that matter, but the jury did make one suggestion: "in time of emergency that unnecessary marine traffic be temporarily halted or diverted so that bridges do not open during rescue operations."[110] Apparently, an ambulance had been delayed when the old span was opened to accommodate marine traffic during the rescue.

The jury then rested and McDonald closed the coroner's inquest to make way for Chief Justice Sherwood Lett's royal commission, which would determine what the jury couldn't: the cause of the collapse.

9

The Royal Commission

Our company appreciates the prompt appointment of a royal
commission to investigate the disaster. No effort will be spared
by the company in co-operating with the commission in
ascertaining the cause or contributing causes.[111]
—ALLAN S. GENTILES, VICE-PRESIDENT AND MANAGER
OF DOMINION BRIDGE'S PACIFIC DIVISION

It was perhaps only a matter of circumstance that Chief Justice Lett found himself being invited to lead the royal commission at all. During World War II he had turned down an offer from Prime Minister Mackenzie King to become a judge of the Supreme Court of Canada, his commitment to his overseas troops being more compelling. That decision, however, irrevocably changed his life when an errant piece of shrapnel smacked into his leg while he was assessing an assault on Louvigny, a village in the Orne River Valley in northern France. This was his second wound of the war, the first sustained when his LCT (Landing Craft Tank) was raked with machine-gun fire during the raid on Dieppe. No longer able to serve at the front, he returned to Vancouver to resume the law career he had set aside at the onset of the war a few years earlier.

But Sherwood Lett was born to lead and within a few years found himself the Honorary Colonel Commandant of the Royal Canadian Infantry Corps. In addition, he continued to serve the military on a contractual basis, having been invited to join an International Supervisory Commission as Canada's representative to monitor an armistice between North Vietnam and French troops to the south. It was while he was in Saigon that he was invited by federal Solicitor General Ralph Campney to replace Wendell Farris as Chief Justice of the Supreme

Chief Justice Sherwood Lett appointed to head the royal commission. PHOTO WHITEFOOT STUDIO, UBC ALUMNI

Court of British Columbia. Farris, Lett's friend, had recently passed away.

This was now the third attempt to conscript Lett into public service. Not only had he spurned Prime Minister King's offer, but he had dodged an early attempt by Campney to entice him to join the Supreme Court of BC under Farris as a judge. Now he was being offered the position of Chief Justice of the Supreme Court of BC. This would likely be the last invitation he would receive, as he explained in a letter from Saigon to his wife Evelyn: "The Chief Justiceship is not offered more than once in a lifetime and if I should turn it down this time, it would be the last time it would be offered."[112]

Lett's biggest regret in accepting the position was that he would have to resign from all the various boards and positions that he had spent a lifetime accruing. And, "as a judge, and particularly as Chief Justice, he had to interpret the law and show no favouritism whatsoever."[113] This would be especially hard when it came to the end of the royal commission when it would be time to prepare his summation and recommendations. Lett's client of more than twenty-five years was none other than the bridge company, Dominion Bridge, and his lifelong friend was the respected consulting engineer, Bill Swan, who had distinguished himself on the field of battle as well.

When Attorney General Bonner had first called Lett with the proposition, an unpaid one at that, Lett had immediately disclosed the fact that Dominion Bridge had been a long-term client. "Would my appointment still be acceptable?" he asked. The attorney general said that he would consult with the premier and get back to him. Within half an hour, Bonner was on the phone telling Lett that if it was okay with him, it was alright with the premier. Lett then offered that he would be prepared to accept on the following conditions: "(1) that the terms of reference should be wide enough to enable me to make a full and complete inquiry as to the cause of the disaster, and (2) that I should appoint my own counsel and be at liberty to call any such advisors as I felt necessary."[114]

To those demands, Bonner acceded, and further offered that he would deliver the pending order-in-council in person. True to his word, he was on Lett's doorstep the following morning, having just arrived from Victoria by plane. It was obvious that the government had an agenda. Premier Bennett had already informed Lett that he could call any experts he felt necessary, Gaglardi had offered to send over his complete correspondence file as well as be available as a witness, and now Bonner had hustled here with the government's simple message: "What part did negligence play in the tragedy?" It was clear that Bennett wanted a speedy conclusion to the commission to appease the public and the government's political detractors, especially with respect to any involvement by the Toll Authority. The language of the order-in-council (No. 1466) made that abundantly clear:

> AND THAT it is deemed advisable in the public interest to appoint a sole Commissioner to inquire into any and all of the circumstances surrounding, leading to or having any causal connection with the aforesaid collapse and specifically to determine what technical or engineering advice the British Columbia Toll Highways and Bridges Authority and any contractor or contractors in any way involved in the construction of the bridge received in connection with its design, erection or construction, whether such advice was sound, and whether such advice was followed or to any extent disregarded by any person or persons in the employ of the British Columbia Toll Highways and Bridges Authority or by anyone acting on its behalf or by any contractor or subcontractor engaged in this undertaking, and to ascertain whether the negligence or faulty judgment of any person, persons, firm or corporation in any way contributed to or caused the said collapse.
>
> AND TO RECOMMEND THAT, pursuant to the authority aforesaid the Honourable Chief Justice Sherwood Lett of the Supreme Court of British Columbia, be appointed a sole Commissioner to inquire into the matters aforesaid and to report thereon in due course to the Lieutenant-Governor in Council [...]
>
> AND THAT the Commissioner be requested to report his findings to the Lieutenant-Governor in Council with the utmost dispatch consistent with the holding of a thorough inquiry into the matters aforesaid.[115]

While Bonner and Lett were discussing the terms of the order, the Attorney

General was interrupted by a call from Victoria. Concluding the call, he informed Lett that the terms of reference of the royal commission were going to be released to the public immediately. The two then reviewed "possible counsel, clerk, secretary and expert engineering adviser."[116] "I'd like Dr. Pratley of Montreal," Lett stated, regarding an engineering advisor. "I think it's best if it's not a local man, one who might be connected with government work, contractors or the Toll Highways and Bridges Authority. Preferably a Canadian,"[117] he concluded. Dr. Philip L. Pratley accepted the appointment and informed Lett that the practical work would be undertaken by his representative in BC, Andrew B. Sanderson, P. Eng., of Victoria. Dr. Pratley had also participated as a design engineer for the royal commission appointed in the wake of the 1907 and 1916 Quebec Bridge collapses where a total of seventy-three men had died. He was also reasonably familiar with the west coast, his firm, Monsarrat and Pratley, having designed the Lions Gate Bridge two decades earlier.

Sanderson wasted no time in visiting the bridge site, reporting immediately to Lett that DB engineers were concerned about additional spans collapsing and piers being pushed over. Span 4 was slipping down Pier 13, wedging it to the north, and if its slump wasn't arrested, the pier might topple with a domino effect. Sanderson had informed the company to take immediate protective measures, action that Lett wholeheartedly agreed with even though the commission's technical appointees had yet to conclude their fieldwork. It was, perhaps, a relief to Lett that DB was seeking permission, a position that was soon confirmed by Allan S. Gentiles, DB's vice-president and manager of the Pacific Division, when he promised his company's full cooperation.

As Lett was organizing the royal commission, the city paused in mourning as families of the victims said goodbye to their loved ones. Parts of some of the funeral notices were succinct: "Died suddenly June 17," they read. At Murray McDonald's service, the Reverend T.M. Badger of Shaughnessy Heights United Church spoke of McDonald's "integrity of character, his concern for others, his devotion to duty and his unselfish service . . . He never spared himself nor his own safety in his devotion to the highest standards of his profession."[118] Lieutenant Colonel Harry H. Minshall perhaps said it best: "He died with his men. Similar citations of skill and courage accompany the award of the Victoria Cross."[119] And in other sad notices, the victims' families were mentioned; children would be without fathers, wives without husbands.

At John McKibbin's service, the first to be held, Reverend R.R. Cunningham warmly praised McKibbin and the other victims to over one hundred people at St. Andrews-Wesley United Church in downtown Vancouver, including twenty-five teachers and the principal from Barbara McKibbin's elementary school. The Reverend Sam McKibbin, John's father, would write a letter to the city a couple of weeks later:

> Words cannot express how we feel for those who, like us, are passing through this lesser Calvary. Our children loved your wonderful land, and stories of the noble beauty of Canada and the gracious friendliness of her people crowded every page of their letters. I hope I am not asking too much to express through your great paper our thanks to all Canadians for their kindness to our son when he was alive and now his stricken young widow.
>
> The words she spoke to us over the long distance telephone yesterday told us of the amazing helpfulness of their so many friends and the kindness of the Dominion Bridge Company. This brought relief to our saddened hearts.[120]

Reverend Cunningham, who also presided over the funeral of Frank Hicklenton, said further that "The tragedy has shocked the whole community, and has brought a wave of sympathy... Sympathy for bridge victims' survivors should not end in emotionalism. Sympathy should be translated into concrete aid..."[121] He then praised *The Province* newspaper for setting up the Fami-

Left to right: Jim Robertson, Murray McDonald, Lieutenant Colonel Harry H. Minshall, Angus McLachlan, Harry Savage, Vic Brait. COURTESY PAUL MCDONALD

lies Fund, which had already attracted thousands of dollars in relief, Allan S. Gentiles personally donating a sizeable $750: almost two months wages for the average ironworker.

Prominent Vancouver citizens stepped forward to offer to administer the fund, which was to be held by the Royal Trust Company. Included in this parade was Lloyd Whalen, president of the Vancouver Labour Council; Benton Brown, former president of the Vancouver Board of Trade; William Manson, former vice-president Pacific Region of the CPR; Edward A. "Teddy" Jamieson, executive officer of the old Trades and Labour Council; and Ralph Baker, president and managing director of the Standard Oil Company of British Columbia.

The relief was welcome given that the Workman's Compensation Act forbade families from launching lawsuits against employers who paid into the WCB fund, DB being a strong contributor. Widows would have to be content with a one-time $100 carry-over allowance followed by a $75 monthly base pension plus $25 per month for each eligible child. This was a quarter of what ironworkers made at the time, forcing many widows to seek work to make ends meet. Sixteen families benefited from the pensions, two of the victims being single. J.P. Berry, WCB solicitor, asserted that although the families couldn't sue DB, they could sue a third party such as Swan, Wooster & Partners, should they be found negligent.

Just before Chief Justice Lett opened the royal commission, he faced his first challenge. The Local 97 executive members were angry that they were being denied their own representative engineer on the commission. The royal commission already had two engineers—this would soon grow to ten—to investigate the technical aspects of the collapse: Dr. Philip Pratley and his associate Andrew Sanderson. But the Local wanted its own man to protect and represent its interests. After writing a letter to the premier and to Chief Justice Lett to request that Lieutenant Colonel Harry H. Minshall, P. Eng., a former DB executive engineer, be allowed to join the technical investigation as a representative of the union, the Local received a letter from John L. Farris, the royal commission's senior counsel. He wrote that "There is no panel of engineers nor are various bodies represented on the investigation . . . However, the commission would be pleased to receive assistance from any competent source. Therefore, if your union wishes to retain the services of Mr. Minshall would you kindly ask him to get in touch with me."[122]

Union vice-president George Sinclair protested that Farris was essentially asking the union to pay for the services of an engineer to aid the commission, a position that he promised to take up in a second letter to Chief Justice Lett. Although Lett met with Minshall, he explained that the commission already had two consulting engineers and that if he wanted to participate he was welcome to, but would have to work out the details of payment with the union.

At 10:30 a.m. on July 9, Lett opened stage one of the inquiry in courtroom no. 414 of the Vancouver Court House, with a twenty-minute preliminary hearing to introduce counsel and the representatives of the various interested parties, and to enter into evidence the dozens of exhibits. The hearing would then adjourn until July 21 at which time the commission would begin stage two, the deposition of the many eyewitnesses. It was only after that, during the third stage of the hearing, that the technical aspects of the collapse would be presented by the various experts.

Introduced were the commission's senior counsel, John L. Farris, QC, whose uncle was Wendell Farris, the former chief justice whom Lett had replaced. Farris's assistant, assistant commission counsel, W.J. Wallace, was then introduced, as well as the commission secretary Robert Wilson, who opened the hearing by reading the order-in-council appointing the chief justice to lead the commission. Lett promised to conduct the commission with "the utmost dispatch consistent with the holding of a thorough inquiry. It will be as thorough as I can make it and I am relying on those who can assist to accord the same degree of co-operation and assistance they have shown so far." He further stated that there might be a further delay to permit the necessary scientific evidence to be gathered and analyzed. "I'm not going to attempt to set any time limits in which these tests should be completed," he concluded.[123]

In addition, George H. Steer, QC, of Edmonton, and Vernon R. Hill, were introduced as counsel for the consulting engineers, Swan, Wooster & Partners; H. Lyle Jestley and W.S. Henson were introduced as senior and assistant counsel for DB; and J.A. Clark, QC, J.S. Maguire and R.C. Bray as counsel for the subcontractors, Kiewit-Raymond. Ralph Sullivan appeared as counsel for WCB, J.E. Boughton as counsel for Lieutenant Colonel Harry H. Minshall, and M.M. McFarlane, QC, and E.J.C. Stewart, as counsel and assistant counsel respectively for the British Columbia Toll Highways and Bridges Authority. Other representatives included Norm Eddison and Herb Macaulay, elected representatives for the Vancouver-New Westminster District Building Trades Council. Macaulay was also representing Local Union No. 138 Brotherhood of Painters, Decorators and Paperhangers of America.

Additional experts would be introduced later. They would include William M. Armstrong, professor of metallurgy, UBC, who would test the steel of the upper grillage as well as examine the connecting plates and perform shear tests on similar bolts to those used to fasten the tie plates connecting Spans 4 and 5;

Major Ross P. Dunbar, Royal Canadian Ordnance Corps, who would examine the steel for signs of explosives; Allan A. Kay, chief engineer of G.S. Eldridge & Co., consulting engineers, who would test the stringers of Bent N4 as well as perform various other tests; Alexander Hrennikoff, professor of civil engineering, UBC, who would investigate the Bent N4 grillage as well as the use of wood as diaphragms and plywood as "softeners;" Robert N. McLellan, president of Robert McLellan & Co. Ltd., who would investigate the 111.1-tonne traveller; and William Pryde, district sales manager, explosive division, Canadian Industries, who would prepare a report on the possibility of explosives being used. It was an impressive list of industry experts, and it reinforced Lett's intent to be thorough.

Conspicuously absent though was Dr. Phillip Pratley who was too ill in Montreal to attend. In fact, he would succumb to his illness on Friday, August 1, 1958, well before his testimony would be required. Between Sanderson, Farris and Lett, it was agreed that a replacement would have to be found and that an American should be appointed in his place. Sanderson submitted a list of names and Lett chose Dr. Frank M. Masters of Modjeski and Masters, one of the oldest bridge engineering firms in America. The firm got its start in 1893, only a decade or so after Dominion Bridge began its legacy of bridge building. Dr. Masters, of Harrisburg, Pennsylvania, joined the firm in 1924, and was now one of the most senior and respected bridge engineers in the United States. He would be accompanied by one of the firm's partners, J.R. Giese.

Simultaneously, other engineers were appointed to assist with the field investigation on behalf of Dr. Masters. Lett conferred with Attorney General Bonner and it was decided to cast the net further afield to seek the appropriate level of engineering expertise to aid the commission.

J.R.H. Otter of the firm Rendel, Palmer & Tritton (now High-Point Rendel), of Westminster, England, was appointed together with Ralph Freeman of Freeman, Fox and Partners (now Hyder Consulting), also of Westminster, England. Rendel, Palmer & Tritton began building bridges in 1822 when imagination and intuition played more of a role in bridge building. Freeman's father was the design engineer of Australia's Sydney Harbour Bridge, circa 1932—the same bridge that John McKibbin and his friend Graeme Kelleher had chosen as the subject for their joint thesis. Freeman (later knighted, as was his father), was the engineer appointed to assist the royal commission. These men were some of the most important bridge engineers in the world: it was judicious that the commission had requested them, and fortunate that they had accepted.

At the end of the hearing, Farris introduced exhibits 1 to 299 in one large file. They represented "plans, records, reports, correspondence, routine tests of cement, steel and the structure during construction, meteorological reports and 111 photographs taken by police and commission staff."[124] The exhibits would aid the commission as well as any of the participating parties who wanted to review them. They would be referred to extensively throughout the third stage of the commission, and would swell by another thirty-nine files before the commission rested.

Two days before the commission resumed, Minister Gaglardi was invited to a Vancouver Real Estate Board dinner at the Hotel Georgia in downtown Vancouver where he talked about the bridge collapse. "I don't care what they do to me—they can saddle me with all the responsibility they like. That doesn't faze me in the least. What concerns me is that all those strapping young men have lost their lives," he said.[125]

Then he proceeded to tell them about one young man whom he had met in the hospital, a man who could only be Gordy Schmidt:

> He was in such good spirits and he kept telling me how lucky he was. He was inside one of the big box girders on the front span when it collapsed. Then he pulled down the covers and showed me his leg had been torn off at the knee. He said he'd tied his belt around the stump and then crawled clear of the water up the inside of the girder. They cut him free with torches and burned his side and shoulder. If that young fellow could go through that and still be in such good spirits, I should be the last man to burden you with what I have to bear . . .
>
> Regardless of what you read in the papers, or hear on the air, the contracting firm and consulting engineers, who are primarily responsible at this stage of the game, and my department and the government, all feel terrible about this. I don't know where any particular blame can be placed at this time.

At the end of his talk, as the delegates were lining up to shake his hand, he stated that "People are very kind to me," and that he was welcoming the official inquiry. "I will cooperate 100 percent with the commissioner. I'll be happy to have all the department's work examined."[126]

On July 21, in courtroom no. 326, Secretary Wilson called the commission to order for the three days that it would take to question the eyewitnesses. Leading the discussion on the first morning was Andrew Sanderson, who had very quickly come up to speed to present a thorough description of the bridge and its construction process in terms that the various non-engineering types could better understand. His prologue was essentially a short course in bridge engineering, but was complex enough that lawyers' heads were seen dipping occasionally toward their engineering and ironworking clients for clarification as he spoke. "Any cantilever bridge with a central span over 1,000 feet is considered a long and important bridge," he noted. "This bridge with its 1,100-foot central span would be rated in the long span class," he added, putting the size and importance of the structure in perspective.[127]

Over the next three days, the commission heard from forty ironworkers, eyewitnesses and various experts. A list of people having knowledge of the disaster had been gleaned from various newspaper reports by commission personnel, and then had been compared to lists of the men working on the bridge that day provided by DB and Boshard. For the men who did appear and who had experienced the collapse first-hand, fact and opinion became somewhat blurred in their retelling of the event. It had all happened so quickly that for some the bridge initially dipped only fifteen centimetres, but for others it was almost a whopping three metres. Where the noise came from was also a matter of judgment. Farris queried Bill Stroud as if he were on trial: "And it is your oath, is it, that the sound you heard came rather from Pier 14 than from N4, is that your oath,"[128] he asked, in a interrogative attempt to have Stroud objectify what was for him merely his opinion. There was little dispute, however, about what was going on before the collapse, and the comments that were made about that period were registered firmly as fact.

Bill Stroud told Farris that in the month of April, the raising gang had been instructed by Mel Alexander to remove the kicker plates between the lower chords of Spans 4 and 5 and install bracings using 8.9-centimetre toggle pins. This was to facilitate the cantilevered span being lowered onto the next set of falsework, Bent N5:

> The holes didn't line up either to a fabricating error, a shop error, but the system wasn't fabricating, so the pins wouldn't go in there. Rather than explain to the engineer and foreman that would be in charge there that the trouble with the holes not lining up we would go down there and there would be one piece lying there and another over there, I joined them together with an inch and a half pin. I was able to put this in the hole that was out of line in these three holes. So when I left that job there they each had an inch and a half pin in these three inch holes.[129]

Stroud was pulled from the job before it was finished and never returned to the site, but did tell the foreman that the pin holes didn't line up and that he had to resort to smaller pins to make the connection. During the shutdown, however, which was not a complete stoppage of work on the bridge, he suggested that the pins might have been changed.

They were, in fact, only being replaced by Bill Moore and Sam Rouegg on the day of the collapse. Moore and Rouegg had already widened the holes and inserted the correct 8.9-centimetre toggle pins on the west side, but had yet to place the jacks under the assembly, though these would not be required at that stage in the construction process. On the east side of the bridge, the toggle pins were still the original 3.8-centimetre pins that Stroud, Atkinson, Glendinning and Chrusch had installed the previous April. They would shear during the collapse while the larger pins would not.

Jim English described for Farris the equipment: the locie and its bogies; the traveller and how it crawled forward with the steel. He also spoke practically about the men he had been working with who didn't survive, and his testimony set the tone for those who followed. There was very little discussion of death. Injuries were glossed over or ignored as Farris burrowed deeper into the technical description of the collapse, occasionally interrupted by the commissioner for clarification or justification. Noise was a big issue for Farris when interviewing Jim English as well and he persisted until it was described and compared: "This loud bang that you heard to the north, could you describe that in any more detail? Did it resemble any other kind of noise that you have ever heard?"[130] he asked. Some of the men had suggested that they heard sounds of an explosion and Farris wanted to get to the bottom of it. *Was it sabotage?* he was likely wondering.

Whether planned or meant to catch English off guard, Farris suddenly changed course from a detailed discussion of the traveller, to what Murray McDonald was like: "You were taking orders from Murray McDonald, is that right?"

"Yes, I was—he was the Resident Engineer and I took advice from him."

"Can you tell us what he was like?"

"Well, he was a very efficient person, one of the most efficient people I had the pleasure to work with."

"Would you describe him as conscientious?"

"Very much so, yes."

"Now, when you were out behind the train—actually the cars, wasn't it— and between it and the derrick, is that when you talked to McDonald?"

"Yes, I had talked to him there."

"And that would be how long before the accident?"

"Oh, ten minutes."

"And did he say anything to you about safety at the time?"

"He asked me how I felt about the job."

"Well, how did you feel?"

"I told him I never felt more confident in my life on any job I worked on."[131]

On the final day of depositions, Charlie Geisser was on the stand when Norm Eddison, appearing for the Vancouver–New Westminster District Building Trades Council, asked him about the competency of the raising gang, including the superintendent, Jim English, who had only recently been asked the same question about Murray McDonald.

"Mr. Geisser, during my long association with you and knowing all that you have done, I have come to highly regard your opinion. I would like to ask you in your opinion what do you think of the degree of competency of the raising gang, the raising gang foreman and the superintendent of the bridge, and that is in comparison to those ironworkers that you have worked with over a period of years in thousands of tons of steel erection?"

"Do you wish my opinion?"

"Yes."

"Well, the raising gang that I had were [a] very good competent bunch of men. The superintendent, Jimmy English, was a very good man. I figured that he was put in the right place . . ."

"Would you say, Mr. Geisser, though, that Jimmy English, the superintendent, in all cases would call the play on the front end, the erection of the steel? I suggest he was there at all times, practically at all times of iron erection, the main members. In your opinion, would his direction be that which was followed if there was a matter of difference of opinion?"

"Well, his direction would be followed, yes, but he would get his directions from Murray McDonald, I imagine, because Murray was in charge of the engineering on the bridge and everything that Murray wanted done Jimmy would more or less have to do it."

"That is true enough. Thanks."[132]

Sam Rouegg and Bill Moore were next. They were both closely examined about what they had talked to McDonald about at the top of Pier 14 just before the collapse. Lyle Jestley, DB's counsel, took the floor after Farris had finished his examination of Bill Moore.

"You have known Murray McDonald some time?" he asked.

"Yes, I have worked on other jobs with him," Moore responded.

"Isn't he known as a bit of a worrywart?" Jestley asked. "Didn't he worry about every job he was on?"

"He was very conscientious,"[133] was all that Moore would provide.

Competency and the possibility of an explosion appeared to be recurrent themes in the examination of the witnesses as stage two of the royal commission wound to a close. Walter Skorodynsky ended the second day with a description of the noise that he had heard the night before the collapse and his intent to tell someone in authority about it the next day. Bill Moore piped up: "May I ask a question?" After no objections were made, Moore asked Skorodynsky:

"Now, you say there was nobody to report this loud noise you heard to?"

"No, not that I know."

"There was a full-time watchman upstairs?"

"I never thought about watchman. I thought about somebody in charge like a foreman or engineer . . ."

"Well, there was a watchman on duty upstairs all the time."

"I know it, but I never think about tell him," he finished.[134]

In fact, Fred Barriball, the night watchman, was on the north bridge approach with his faithful mutt, Patty, and although he must have heard the same three cracks, he also didn't report it.

On the last day of stage two, the commission heard from Arthur Francis, director of accident prevention, and Walter Miller, safety inspector, both of whom worked for WCB. Francis testified that there was no regulation for a safetyman on the bridge, a statement that Wallace, who was examining, confirmed with Miller. W. Henson, DB's solicitor, cross-examined Miller with the same question. Everyone, it seemed, wanted to be reassured that what they were hearing was correct: WCB had no regulation regarding a safetyman on the project? Despite that, when Wallace asked, "what was your impression in so far as the safety conditions on this particular job were concerned?" Miller satisfied the commission with the lukewarm response, "Well, I think it was a good job."[135]

At the conclusion of the final day of testimony, Commissioner Lett recessed the commission until further notice. Two days later, on July 23, a meeting was held in his chambers with commission participants to determine when the commission should resume. Colonel Swan wanted to reconvene immediately while Farris suggested two months would be adequate for the technical detail to complete their studies. In the end, Lett selected September 30 as the opening of the third and final stage of the inquiry. The various engineers would spend the rest of the summer pouring over metallurgical samples of the steel, testing the shear strength of the bolts, closely examining the upper grillage, reviewing the last movements of the traveller, examining Piers 13 and 14, and reviewing the use of wood diaphragms in the upper grillage and plywood as "softeners."

During the meeting, John Prescott mentioned that DB was starting shop work on the south-end steel, and still hoped to meet the June 30, 1959, forecast completion date as the company had been six months ahead of schedule at the time of the collapse. He asked for permission to start salvaging Traveller No. 1, but Lett told him to hang on for a bit, that he would inform him when the company could begin salvage work. No more than a day later, however, he gave the go-ahead by announcing that DB could commence salvaging the traveller on August 1, as Otter and Freeman, the two English consulting engineers, would be finished the physical part of their investigation by then.

Although the steel would not be formally released back to DB until October 17, the company needed the traveller to facilitate steel erection on the south shore. Assisting the recovery was a barge with a small crane mounted on it, to be replaced within ten days by Traveller No. 3 mounted on two barges. Also aiding the salvage was an underwater camera and top-side television screen to make sure that the divers were cautious about how they untangled the equipment from the steel. Unknown stresses, once released, could be potentially fatal.

Watching every step of the process were curious onlookers and perhaps not surprisingly, many tourists. As far away as Los Angeles, California, tourist bureaus were telling their clients that they would never see a sight like this again, and on the old Second Narrows Bridge traffic slowed to a crawl as sightseers craned to catch a glimpse of the crumpled bridge. Although this had the potential of putting more pressure on the Lions Gate Bridge and the North Vancouver Ferries, the latter of which the City of North Vancouver had just voted to cease altogether due to lack of provincial funding, traffic flowed surprisingly well even after the ferry service stopped.

Less than a month later, Chief Justice Sherwood Lett received some rather disquieting news in the form of a rather ominous four-page airmail letter from Ralph Freeman, who had only recently returned to London after his inspection of the bridge. It was marked "secret," and its introductory sentences referenced its importance: "I am sending this letter to you under secret cover, for your eyes only, because premature leakage of the matter I wish to put to you might conceivably prejudice the course of the inquiry . . . The matter to which I refer is sabotage."[136]

Although Freeman did not really believe that the cause of the collapse was sabotage, he was concerned that DB might use any and all means to mask their design errors, which Freeman had identified as failure to install web stiffeners on the upper grillage and lack of effective lateral restraint on the top flanges of the upper grillage. The first would have prevented the beams from "bulging sideways," while the latter would have prevented them from "folding sideways." He was fairly caustic in his remarks and felt that the design fault . . .

> . . . could very seriously damage the Dominion Bridge Company's reputation. I believe the injury would be less severe and less permanent if they were voluntarily to admit their mistake and squarely face the consequences. But I fear that their Directors will not take this line and that their Insurers will not allow them to do so. They will insist that they seize on anything calculated to divert attention from the design errors and seek to prove some less damaging explanation of the collapse.[137]

Barring proof of earthquake, excessive wind, a boat collision or facetiously, even a plummeting meteorite, Freeman wrote, sabotage was the next most plausible defence if it could be proven.

The flurry of letters that flew back and forth across the Atlantic suggested that there was a genuine interest in this line of thinking. Others were pondering the same thing. Minshall, who had worked out a satisfactory arrangement with the ironworkers, admitted in a meeting with Lett that it had crossed his mind, but that he had heard of no discontent among the workers. And besides, an explosive device that exact—emanating from the top of Pier 14 and just strong enough to blow the bolts on the tie plates—could only have been set by a professional.

Not too unexpectedly, the bridge company did try to manoeuvre the actual cause away from the falsework. On August 30, DB informed Farris that "they felt they would be able to satisfy the commission as to its (falsework) construction and they were now looking elsewhere with a view to determining the actual cause of the collapse."[138] Farris informed Lett that the company was now convinced that the problem emanated from the top of Pier 14 with the tie-down bars and "soft blocking" consisting of 4 WF I-beams, steel packing and small timbers, which they felt had loosened due to the repetitive action of the locie, but they asked that this information be withheld from Freeman and Otter until they had time to study it further. They would know more once they had salvaged the bent and grillage, but an argument between DB and the Toll Authority over who should pay for the damaged piers was delaying that from happening.

The crushed upper grillage with the wooden diaphragms still in place. OTTO LANDAUER OF LEONARD FRANK PHOTOS, JEWISH MUSEUM ARCHIVES LF-35515

DB insured everything through the Toronto General Insurance Company which, according to John Prescott, covered the company "against damage to property of others," but not damage to DB's own property. The insurance company refused to pay, arguing that as soon as the company had rested its steel on Pier 14 it belonged to them. In fact, the pier, once complete, was owned by the provincial government. The Toll Authority was now suing DB for the cost of repairing and replacing the piers, thus instigating a suit by DB against the insurance company. Although DB was successful in their claim, it was later overturned on appeal. Before the issue was resolved, however, salvage of the bent and grillage assembly would be delayed.

The delay, plus the pre-collapse construction deaths and then the bridge collapse and resultant loss of life, was adding strength to the myth that a Native curse surrounding the small gravel islet, Hwa-Hwoi Hwoi, located exactly where the shattered Pier 14 now rose from the inlet Tower-of-Pisa-like, was continuing to jinx the site. Hwa-Hwoi Hwoi had been removed in 1923 to provide fill for the Ballantyne Pier across the harbour, but this was after a medicine mask from a strange tribe had been found on the island, causing local Natives to shun the site.

Construction of the south-side steel also would be delayed. The company had salvaged Traveller No. 1, which was sitting at the plant, but some of the special alloy steel parts required for its reconstruction were not available in Vancouver, which necessitated accessing eastern mills to manufacture these custom items. Unfortunately, the eastern mills were on strike with no resolution in sight, something that the ironworkers were respectful of and ironically would soon be experiencing themselves.

The delays were, perhaps, responsible for heating up the ongoing debate about another crossing. Minister Gaglardi advised that City of Vancouver engineers were doing traffic studies and that his department was studying it as well. "When reports are completed, both sides—the municipalities and the provincial government–will meet again," he said. "That is expected to be in about three months."[139]

On September 17, Dr. Masters submitted his commission report and sent a copy to DB. It took a few days for the company to read it, absorb it and react to it, but five days later, DB lawyer Lyle Jestley phoned Farris, fuming. Jestley charged that although DB Pacific manager A.S. Gentiles had admitted that "They made a mistake and they know it,"[140] Dr. Masters had made some unfair comments in his report, which pointed to the upper grillage as the cause of the collapse. Lett made notes of Farris's discussion with Jestley in his diary: "They suggest that if it be assumed that the fault lay in N4 still there was no want of attention paid to the design of the grillage. If it was in error then it was an error in judgment. It had been checked, but no one found it and, therefore, it was not negligence."[141] An error in judgment was easier to swallow than the crippling word "negligence," but the commission was soon to hear that someone had, indeed, found one of the errors prior to the collapse, but had failed to report it.

On the same day that Dr. Masters submitted his report, Farris wrote to Jestley on another matter: "It would be extremely helpful if you would inform me prior to the hearing who you will be calling as witnesses, and the nature of the evidence they will give."[142] When DB did respond, noticeably absent from the list of potential witnesses were A.S. Gentiles, vice-president and manager of the Pacific Division; Angus McLachlan, chief engineer; Cliff Paul, the third engineer assigned to the bridge; High Carpenter, the general superintendent; and a host of others who had intimate knowledge of the construction project. A surprise inclusion, however, was Robert Eadie, vice-president and manager of the Eastern Division, a man who had significant bridge-building experience, but little knowledge of this specific structure.

McLachlan didn't appear because he was still reeling from the shock of losing so many men, especially his friend and colleague Murray McDonald. According to John Prescott, "Dominion Bridge wouldn't let him get on the stand because he was so upset that . . . they didn't know what he was going to say." McLachlan, born of an age and disposition where honesty and integrity were the mark of a man, was planning to take full responsibility for the debacle, according to his son. He didn't want the reputations of his friend and his friend's assistant impugned. DB would have none of it, so kept him off the stand. "I think it killed him . . . he was never the same after that," his son would later comment. The loss gnawed at McLachlan throughout the remaining seven years of his life, probably hastening his demise from cancer just before his sixty-seventh birthday.

On September 26, only five days before the royal commission would reconvene, Lett, the commission consultants and various representative counsels, met to map out a strategy. The chief justice, after consideration, agreed that if DB and Swan-Wooster were in support of the commission findings, then a unanimous report signed by all of them would be prepared and be available to be examined—and if necessary cross-examined—during the third stage of the proceedings. This would mean that DB had to come to grips with the fact that the cause of the collapse was the poorly designed upper grillage of Bent N4, which they were now, reluctantly, willing to do. A few days earlier, Swan-Wooster had come to the same conclusion after reading the commission expert's findings.

DB's acceptance of the commission's findings, however, came with a price. Lett wrote about it in his diary a day before the commission reconvened:

> Mr. Jestley stated that the Dominion Bridge would admit responsibility, and hoped it would not be necessary to bring in the names of Messrs McDonald and McKibbon [sic] deceased. Mr. Farris stated he felt it would be necessary for him to bring out in evidence the name of those who prepared the design sheets for Falsebent N4, and to determine by evidence whose responsibility it was for checking same, and that he would have to ask these questions. Mr. Jestley stated that he would object to such questions being asked, and it was up to the Commission to rule on the matter. I stated that my present feeling on the evidence so far was that it was unfair to place blame by way of negligence or faulty judgment on any particular individual if the evidence showed there was a divided responsibility between various officials of the Bridge Company. However, I felt that if the names were not brought out in evidence of those responsible the public would wish to know, and no doubt the newspapers would make their own examination of the exhibits. I further stated that I felt the Dominion Bridge was a responsible legal entity, and must take responsibility for the acts and omissions of its servants in the course of their employment, and that my present feeling is that a finding of responsibility could only be fairly made against the Company and it would not be necessary to name individuals in my report."[143]

DB had no choice but to accept the decision of the commissioner who could no longer show any loyalty to the bridge company that he had served for two and a half decades. Within a few weeks, his summation would make that abundantly clear.

10

The Royal Commission: Stage Three

*In bridge building, as in any other hazardous occupation, the
human element plays a very important role and for the success
of any scientific achievement, where trade and profession are
practiced, these two endeavours must, of necessity, become
co-existent. Factors involving the application of human effort
extend beyond the end of the slide rule or electric computer
and, when coupled with scientific knowledge, become known
as the factor of "calculated risk." It is in this bracket that the
competency of fellowman plays such an important role in our
achievement. In the pursuit of surgery and aeronautics, as in
engineering and many other scientific fields, judgment, the
antithesis of man's formal knowledge, prevails and proves the
adage, "The only world a man truly knows is the world created
by his senses." It is from this understanding that humility
becomes of singular importance in life, ever remembering that
it is by the Grace of God that we are saved from disaster and
that disaster is never an Act of God. The Second Narrows
Bridge disaster, from the data at hand, is, in our opinion,
accidental to human knowledge.*[144]

–LIEUTENANT COLONEL HARRY H. MINSHALL, P. ENG.

The commission reconvened on September 30 with a continuation of testimony from eyewitnesses who were too injured to appear during stage two. Wesley
Ash, a Boshard employee, described how he jumped approximately sixty metres
into the inlet from Span 4 after Span 5 went down, and Michael Joseffy, another
Boshard man, recalled how he rode the collapsing steel of Span 4 into the wa-

The collapsed Spans 4 and 5 viewed from the adjacent wooden trestle. PHOTO BILL HADLEY

ter. Bill Wright, a member of the hooking-up gang, was hooking up the bottom chord, which was still on the bogies, when it dropped out from beneath his feet. Lou Lessard, still wearing a leg cast, described how he quickly lost his balance after hearing a noise.

"You mentioned a noise. What kind of noise was it?" Farris asked.

"Well, something very hard to answer. We can say that is a bolt or pin or this or that. That is steel giving up or breaking. I can't give you the definition. It was quite a ways, about 350, 400 feet from the point where I was so . . . pretty hard to say exactly," he replied in his thick Québécois accent.[145]

Following Lessard's testimony, examination of the various experts began. Robert McLellan, who had been tasked with investigating the traveller, was first. He determined that it and its ropes had been in good condition prior to the collapse and that all of the evident damage was sustained during the collapse. Next, Louis Osopov described how he and his colleagues had been contracted to investigate Piers 13 and 14. In summary, he concluded that the two piers had been built according to their specifications and that they were strong enough to support the loads that had been designed for them. Damage to Piers 13 and 14, he said, was the direct result of the collapse of the steel superstructure.

Following Osopov, Farris questioned Allan Kay regarding his investigation of Floor Stringer Nos. 2 and 3 of the upper grillage assembly upon which he had conducted physical and metallurgical tests. Except for the height variation of Beam No. 2, which was in excess of accepted tolerances, he found that the beams were generally in accordance with American Society for Testing Materials standards. "The high spots of Beam No. 2 may have given rise to irregular distribution of loading on the beam assembly," he said, before also informing the commission that sulphide tests indicated that the beams may not have conformed "completely with the mechanical properties as required by ASTM standard A7."[146]

"Although the sulphide segregation is not too extensive in the specimens examined," he added, "erratic distribution of sulphur could give rise to much more severe local segregation in other parts of the beams": a problem, given that an undesirable concentration of sulphur could weaken the steel.[147]

William Armstrong, the UBC professor of metallurgy who assisted Kay with his tests, corroborated Kay's statements but suggested that although the beams only "partially" met ASTM standards, they were normal for that type of structural grade metal. Where the beams were deficient, he noted, was in their other mechanical properties. Parts of Beam No. 2, which was from a different heat (the other three beams were manufactured together in a separate heat and had high yield points and high tensile strength), had a lower yield point and lower tensile strength than specification required, the yield point being where "sudden plastic deformation begins in this type of structural steel."[148]

When cross-examined by Vernon Hill, counsel for Swan-Wooster, about how a beam like No. 2 could have escaped attention before being used, Armstrong responded that although the quality of the beams used in the bridge was determined by G.S. Eldridge, they didn't perform mill tests on the steel or inspect each beam individually. They relied on mill test reports (Mill and Test Certificates) from the manufacturer who performed varied tensile tests with samples normally taken from the centre of the upper flanges of each beam. Although the sample from the upper flanges of Beam No. 2 had been within specification, other parts of the upper flanges were not. Beam No. 2 had a rolling groove that according to Armstrong "could lower the resistance of the top flange to plastic hinging."[149] Unfortunately, this was also the beam that was taller than the other three, and would therefore have borne more of the load.

Professor Hrennikoff, who had been contracted to do a stress analysis of the grillage assembly, found that the lower grillage was satisfactory, but that the webs of the upper grillage, or floor stringers, were weak with respect to buckling strength. "Now, if stabilizing effect of these diaphragms is ignored completely," he went on, referring to the wooden blocks between the flanges, "then the stringers are grossly under-designed as far as buckling of the web is concerned. The stress in those stringers would be pretty nearly 50 percent in excess of failing stress."[150]

With respect to the wooden diaphragms, he believed that they had some stabilizing effect, but one that would have been hard to calculate. He noted that they were installed about a metre from the centre of the column (legs of the bent), they were on one side only and they were made of wood: "wood is not a satisfactory material for this kind of work."[151]

Another factor he suggested was critical, was that although engineers design for allowable or permitted stresses to be lower than stresses producing failure as a buffer for safety, there was a significant gap in the allowable compression stresses permitted by the American standard, AASHO, over that of the Canadian standard, CSA. They differed by over 30 percent at some level, CSA being more tolerant, and according to Hrennikoff, actually unsafe. He wondered, in fact, why the collective wisdom of the two countries couldn't have addressed this obvious discrepancy. On the other hand, with respect to the design of webs of stringers in buckling, he also admitted a disturbing fact that "There is no definite well substantiated theory available with regard to this aspect of design, which would make proper allowance for all significant factors involved."[152]

In his conclusion he wrote that "Although the designer might have been safer and more prudent to follow the stricter AASHO formula the commission of error in judgment on his part, in the uncertainties of the problem, was made so much easier by false sense of relative security conferred by the CSA formula."[153] He also stated that the designer was under no obligation to follow either standard given that the falsework would not become part of the completed bridge.

Hrennikoff elaborated in a subsequent article:

> Although the use of wooden blocking in lieu of rigid diaphragms was a grave mistake of design it does not fully explain the failure. The allow-able compression stress in the webs of the stringers using the column formula of CSA, 1952 specifications was 8.4 kips/sq. in. and with an additional 33% allowance, common for the erection conditions, 11.2 kips/sq. in., while the actual stress at failure was 17.2 kips/sq. in. The spread between these figures does not seem to be great enough to ex-ceed the usual factor of safety especially if some positive strengthen-ing effect is attributed to the wooden blocks, weak as they might have been. This reasoning points on the one hand to the presence of some additional aggravating factors contributory to the failure, and on the other hand to a positive weakness of the Canadian column formula. Both these conjectures are believed to be correct. The additional factors which influence the failure were:
>
> 1) The action of the plywood packing at the top and bottom flang-es of the stringers.
>
> 2) The imperfections of the shape of the stringers.
>
> 3) The difference of the stringers in depth.[154]

When he was finished, Farris asked him, "A question may be asked how an experi-enced designer could commit an error of judgment attributed to him?"

"I suggest experience may be dangerous under certain conditions," the pro-fessor answered, as he went on to explain that the use of falseworks on other suc-cessful bridge erections was a "vindication of the method of design."[155]

Throughout the rest of the afternoon and most of the next day, Dr. Pratley's representative Andrew Sanderson presented his report, bringing into sharp fo-cus why the royal commission was sitting: of the seventy-nine men working on the bridge that day, eighteen had died and twenty had sustained serious injury. His examination outlined conditions at the time of the collapse, a description of the collapsed structure, an examination of the calculations to determine loads and stresses, and in the end, his opinion based on the gathered facts of why the structure had failed. He also put forward the possibility that if the "soft blocking" and wedges in the jacking chambers of Pier 14 had been dislodged by the movement of the locie, then the floor stringers (upper grillage) of Bent N4, which were already vulnerable, could have failed when there was a change in stress—a position that DB was wholly partial to. But then he quoted a reference from Professor Hrennikoff's report that suggested that "analysis of the stringers in the light of information obtained from various tests performed indicate that the stringers were in a highly critical state and that failure was not only likely but almost inevitable."[156] It was as detailed a report as he had given at the beginning of the commission, and it had reporters in the gallery scribbling furiously in their notebooks.

The following morning, the *Vancouver Province* newspaper headline summarized it neatly:

WEAK GIRDER LED TO BRIDGE COLLAPSE

The Second Narrows Bridge collapsed on June 17 because an "inadequate" support buckled under the half constructed span reaching over Burrard Inlet, the royal commission inquiry was told Tuesday.[157]

In a few short paragraphs, what had been suspected from the beginning, but what had taken weeks to reveal, was now official. Ralph Freeman confirmed it when he read from the introduction and summary of the joint report at the beginning of his examination:

We, the Engineering Consultants and Advisors appointed by the Commissioner, having completed our investigations into the collapse during erection and now present this joint account of our work.

Our conclusion as to the cause appears in paragraph 4:30 of this report and is as follows: The primary cause of the accident is elastic instability of the webs of the stringer beams of the N4 grillage ... accentuated by the plywood packings above and below the beams. The instability was due to the omission of stiffeners and effective diaphragming in the grillage, and this in turn was basically due to an error in the calculations. Such diaphragming as was provided was inadequate.[158]

He went on to explain the term "elastic instability":

Now the behaviour of a strut or column made of elastic material such as steel is fairly well understood. It was enunciated a long time ago in theory by a German called Euler. Since Euler's time—I think that is something like two hundred years ago—many, many experiments and very much theoretical work has been done on the performance of struts. But Euler started all this by deducing a mathematical formula for the critical load or load which would produce failure of what he called a perfect strut of elastic material . . . The condition of the strut at this point when failure takes place is called the condition of elastic instability.[159]

The Euler formula is a tool used by engineers to calculate the load-carrying and deflection characteristics of beams. Developed in 1750 by Leonhard Euler, a Swiss (not German) mathematician, the formula wasn't extensively employed until the late nineteenth century when the Eiffel Tower was being built. Since then, it has been especially important to civil and mechanical engineers. Although Freeman explained the theory and then, at Farris's invitation, proceeded to deconstruct the critical worksheet where McKibbin had made his calculations, at no time did Farris ask, nor Freeman speculate, why the two engineers had not chosen a more complex method to calculate the columnar strength of the four upper grillage beams, perhaps accepting Hrennikoff's suggestion that the CSA formula had presented a "false sense of relative security."

Toward the end of the day, Farris caused a stir when he asked Freeman whether the inadequacy of the upper grillage would have been visually evident to an experienced engineer. Hill, counsel to Swan-Wooster, objected to the question, but after a short recess, allowed that it could proceed. Freeman proposed that an erection engineer with little to no design experience might not have noticed anything out of the ordinary. Otter, when questioned next, agreed that it might have escaped his attention too, and Giese suggested that he might not have found it. Even Colonel Swan, who had ample opportunity to view it during erection, when questioned later, said that given his previous experience with DB, he had given it very little thought. "I don't think I would have noticed it," he said.[160]

Each of these men was also asked in turn about the wooden diaphragms. What purpose did they serve? They all agreed that the blocks weren't placed there as stiffeners, but rather as diaphragms, spacers or separators to keep the beams in place, a kind of restraint to keep them upright. They also heard that during the heat of the summer the blocks had each shrunk away from the steel, supplying little, if any, of the support that had been intended of them.

Farris began Dr. Frank Masters's examination early the next morning. He too was asked whether a visual examination would have identified a problem. Admitting that he was an old man, albeit an experienced one, he speculated that he might have seen it whereas a younger engineer would not have. Cross-examined by Lyle Jestley, he was asked, "Now, would you be prepared to say [the] triggering cause of this was the grillage and not some other thing?" Stating that nobody could make that assertion, he went on to qualify his opinion. For the twenty to thirty minutes that the bridge had a full load, essentially right after Gordy MacLean had delivered the bottom west-side chord to the front end, the metal in the upper grillage I-beams was in a plastic state, a process that was slowly altering its structure. At that point, while it was vulnerable, any other source of added stress might have started the reaction. "It was the upper tier of grillage that collapsed and fell, but whether some other thing started it I don't know, I don't think there is anybody [who] can say that," Masters said.

Somewhat predictably, Jestley added, "Well, that is the point I make. You don't say that—you can't say that this really caused it?" Under Farris's redirect, Dr. Masters admitted that if the grillage had been designed with the proper security tolerances, including steel stiffeners and diaphragms, then he could think of no other reason why the bridge would have collapsed.[161]

Later that afternoon, Jestley asked that Lieutenant Colonel Harry H. Minshall be invited to proceed with his report, one that supported DB's argument. Minshall concluded his report:

> Based on all data presently available we are of the opinion that the sudden release of stresses in the vicinity of [Pier] 14 pier top "triggered" or activated the collapse of the most highly stressed erection member, namely, the upper grillage of false Bent N4, causing Span 5 to fall which, during or near the end of its fall, pulled support from under the south end of Span 4, and caused its subsequent collapse.[162]

Although Minshall generally agreed with the findings of the other experts, he believed that the upper grillage had not yet reached the limit of its bearing capacity:

> Knowing these men and having worked with some of them who were present at the bridge for at least twenty years, I am of the opinion that the thoroughness with which Murray McDonald did his work, his ability, he is . . . the most conscientious individual. It is impossible for me to—knowing Murray and other men on the job, that when he came

out and looked at this pier, that had there been evidence that structural failure, deformation in the beams, I believe that they would have been a warning to him and that he would have, in normal fashion, have phoned the office, as Dr. Masters explained, and said, "This does not look right to me." I feel that the beams did not quite reach that stage.[163]

Minshall believed that the sudden release of stresses at the top of Pier 14 was possibly influenced by a change in temperature. Although Span 5 had been tethered to Span 4 with tie plates in April and had been free to expand and contract with the change in temperature throughout the day as is the nature of steel, on or about June 11 or 12, the main tie-down bars in Pier 14 were blocked in position in anticipation of landing the span, "thereby restricting the free movement of the north bearing of Span 5." Then on June 17 another 181 tonnes was added to the end of Span 5, increasing the compression between the bottom chords of Span 4 and 5 and simultaneously influencing the "soft blocking," which was already "set up" by the temperature rise in Span 4.

Minshall speculated on what happened next:

> Should this resistance to the compressive force be suddenly removed, the following load change would occur. The resulting release of stored elastic energy would accelerate and tend to move all portions of Span 5 towards the north. This energy or movement would be transmitted down through the false bent at N4, in proportion to its stiffness, to the top of the upper grillage where it would be manifest as a horizontal impact force toward the north.[164]

Farris's failure to examine or comment on Minshall's theory, was more than merely dismissive. "Mr. Minshall," he asked, "do I understand from this report—I will put it this way—would you agree that there were errors in the calculations that were made in respect of the upper grillage of Bent N4?"[165]

Minshall responded that he had never seen the DB calculation sheet in question and that he had performed the relevant calculations himself.

Farris tried a different tact. "Would you agree that there was elastic instability in the webs of the stringer beams of the N4 grillage?"

"Yes," Minshall replied, concluding his testimony and Farris's examination.

It was now Lyle Jestley's turn to examine Robert Eadie, vice-president and manager of DB's Eastern Division, a man who had little knowledge of the bridge, but who was fully familiar with the practices of the large engineering firm. After

R.S. Eadie, VP of Dominion Bridge's Eastern Division, arrives in Vancouver to testify at the royal commission.
COURTESY SANDRA EADIE

quickly ascertaining that he was in agreement with the joint findings of the commission, Jestley deferred to Farris for Eadie's cross-examination where Farris asked him, again, to confirm the cause of the collapse. Responding once more that he agreed with the commission's findings, Farris moved on to the system that the company had for preventing such errors. He was informed that the calculation sheet, as was the practice, was prepared in the erection department by a capable engineer preparing it and a senior engineer checking it: essentially, John McKibbin and Murray McDonald.

"Now, who prepared these calculations and who checked them?" Farris asked, the suddenness of the question shocking some of the commission participants as well as a specific few in the public gallery.

Jestley, who already had a testy relationship with Farris, telling the commissioner at one point that he had received "scant courtesy" from commission counsel, bounced aggressively to his feet: "Mr. Commissioner, I would advise the witness not to answer that question . . . The fact is that these employees are dead, they are not here to give evidence. We are not sure this is their last calculation. We have assumed it is so and I fear that it is only fair to them under all the circumstances that their names be not given . . . I think nothing further could be served by bringing out the names of these men."

After Farris advised that he regretted having to ask the question, but felt it necessary, the chief justice stepped in: "Well, the terms of the Commission under which I am acting require me to ascertain whether the negligence or faulty judgment of any person, persons, firm or corporation, in any way contributed or caused the said collapse."

Jestley, sensing an opportunity to shift the fault away from his client, piped up: "Assuming, your lordship, the error is an error in judgment or faulty judgment, that is surely the error of the employees of the company, is it not?"

"Well, that may be your submission," Lett replied, curtly. "I would give no ruling on that at this stage, not having heard all the evidence in the matter."[166]

Murray McDonald's widow Mary, who was in the public gallery, was appalled that her husband's name as well as that of John McKibbin's would be made public. Although it was not Mary's nature to protest, it was very much the character of her eldest daughter Jean, who telephoned Sherwood Lett later that day and was advised that if the names weren't revealed during the royal commission, the newspapers would eventually find them out anyway, and likely with much more fanfare and controversy. Later that week, Mary and Barbara McKibbin steeled their nerves to visit the commissioner in his chambers to add their voices of protest over the naming. Barbara remembered that Lett was very kind and thoughtful, but their protest had no influence on his decision. The experience was disheartening, and doubly so when the Vancouver School Board docked her half a day's pay for her excursion.

Farris then shifted his focus to DB's organizational chart and stopped at the chief engineer's job. Virtually answering his own question, he recognized that although the chief's job would have him responsible for all designs, he would need to delegate some of the work. Farris addressed the structural engineer next.

"But did the structural engineer or anybody in his department apply their minds to the design of this upper grillage and Bent N4?" he asked.

"No, because it was believed that two competent engineers who had had experience in the design could carry out the work and check," Eadie replied just before the commission adjourned for the day.[167]

At 10:15 the following morning, Friday, October 3, Eadie again took the stand, and the examination adopted a much more sombre tone under Jestley's lead.

"Mr. Eadie," Jestley began, "would you give the Commission the qualifications of John McKibbin."

John McKibbin was a "well qualified junior engineer capable of carrying out the duties required . . . He had graduated with high standing from the University of Sydney, as stated, and our experience with him while he was with us indicated that he was a reliable designer," Eadie replied.

"Now, would you please give the Commissioner the qualifications of J.A. Murray McDonald?" Jestley continued.

Eadie described McDonald as a much more experienced engineer who had been with the company for twenty-one years, in the erection department for twelve of those years, and who was the second-most experienced erection engineer across Canada. "Let me restate that. He was an engineer with the second most experience in erection problems and we considered him a very competent engineer," he finished, responding to Jestley's subsequent query about McDonald's competency. [168]

Farris, assuming the floor from Jestley, resumed his progress through the organizational chart which had been cut short at the close of the previous day's proceedings, abandoning that course only after establishing that McDonald "was exercising the functions of a field engineer and erection engineer and of a design engineer." [169] Eadie affirmed that the design part of McDonald's job pertained only to the subject bridge, and not to the company as a whole. When asked whether that change, which was not identified on the organizational chart, was contained in any written communication, Eadie volunteered that it was only verbal, but that it was the company's common practice throughout the country.

A few minutes later, Eadie was asked about who made the pencilled correction of one of the dimensional errors on the upper grillage design sheet. Someone had discovered one of the errors (flange thickness substituted for web thickness), astonishingly, before the collapse, and although they had put a pencil mark through the incorrect figure and noted the correct web thickness, they had not followed through with the rest of the calculation or made anyone else aware of their discovery. What was even more astounding was that the grillage calculation sheet likely never left the small mobile office on the north bridge approach. Access was therefore limited to a very few individuals. Whoever found that error, whether out of fear, embarrassment, or both, must have recognized its implications, but for whatever reason chose to remain silent.

Before he could answer, Jestley asserted that he doubted Eadie would know that. Farris agreed, but stated that since DB had chosen Eadie as the company's representative on the commission and not someone from the Vancouver office, then he expected that he would have obtained whatever knowledge was pertinent to the collapse from the local officials.

"Do you know who made the discovery?" he repeated.

"No," Eadie responded.

"That is of your own personal knowledge. Now, I ask you the next question: Have you been informed by officials of your company—do they know as to who made the discovery?" Farris asked.

"Nobody knows definitely who made that note in pencil," Eadie replied.

"Before I pursue that, do they know when it was made?" Farris asked.

"No," Eadie responded.

"Am I right in assuming that when this chart was discovered or located after the collapse of the bridge that notation was on there?" Farris asked.

"The notation was there the day after the accident," Eadie replied.

"And no information as to whether it was put on just immediately it was discovered or prior?" Farris asked.

"We know it was put on prior to the accident," Eadie responded.

In fact, it was only discovered in a filing cabinet in the small engineering office on the north bridge approach after the collapse. The author of that small but significant change would remain anonymous, but Eadie hinted that maybe DB did know something about the identity of the man when he used the phrase, "Nobody knows definitely," and then advised Farris that the company knew it was put on prior to the accident, two hints that Farris didn't pursue.

When questioned by Farris about how an engineer could have made such a dimensional error in the first place, Eadie could only answer, "All I can say, your honour, is that he made what I call a human error, human errors which are inexplicable."

The conversation next flowed to the plans and how the absence of stiffeners could have been missed by all the departments viewing them. Eadie replied that the men in the erection and detail departments were not engineers, but draughtsmen.

"Now, were these plans submitted to the consulting engineer?" Farris asked.

After answering that he didn't think they were, Farris directed Eadie to the terms of the contract that stated falsework plans were to be submitted to Swan-Wooster.

"Now, why were those plans not submitted to the consulting engineer?" he asked.

"Purely an oversight, as far as I can find out," Eadie replied.[170]

Farris then moved on to the responsibility to submit the falsework plans to the consulting engineer to which Eadie replied that it was the practice to submit plans for the erection scheme, but since the falsework was considered to be a piece of erection equipment and not part of the permanent structure, the consultant would not be responsible for it. Neither had the consulting engineer viewed the plans for the first three falseworks which were constructed of Bailey bridge parts. The AASHO didn't call for it and in Eadie's eastern experience, he had never heard of falseworks being the responsibility of consulting engineers except in very special circumstances.

Colonel Swan was next and several minutes were spent on his qualifications: a doctorate in science from UBC, two prestigious engineering awards, an impressive and decorated war record spanning two world wars, and five decades of bridge-building experience. When asked about whether he was satisfied with the bridge company's execution of his plans, he responded that their shop work was excellent and that they had submitted all of their shop drawings to his firm. "We had a staff of from four to six continuously on the checking of those shop drawings, 320 in number, and we found . . . I don't think we found any errors throughout any of it," he recalled.[171]

Hill, Swan-Wooster's counsel, then questioned him about the requirement for the contractor (DB) to submit falsework plans to the consulting engineer, a condition of the specifications that were prepared by Swan's own firm (acting as an agent of the provincial government). After admitting that he knew about the clause, having read it after it was prepared, Swan advised that checking the falsework plans wasn't required under the AASHO, nor was it required by the CSA. Regardless, Swan advised that he hadn't seen any plans for the N4 or N5 falseworks, partly because the detailed plans weren't ready and partly because it was not the practice of contractors to submit them:

> I think I must say that after the pile support had been driven I gave very little thought to the question of the distribution and the erection of the support to carry the load under N4 . . . Dominion Bridge company has been working for me for twenty-five years, they have never had any failure in false work during that period and it was—I wouldn't call it an elementary, but it was pretty much a routine matter of designing the distribution system to carry the required load at N4 to the pile foundation.[172]

Jestley's turn to cross-examine Swan immediately brought back a painful memory for the elderly engineer. When asked about DB's safety record, Swan said, "Well, I lost my son on the Pattullo Bridge, but it was no fault of the bridge company. It was some oil on a girder, on a floor beam and he walked across it and fell into the river."

"I realize it is a painful subject to you, but what you are saying, Col. Swan, is, sometimes these accidents happen in spite of everything?"

"That is quite true," Swan replied.[173]

Jestley then reviewed old ground by asking Swan again about whether a visual inspection would have identified the problem. Swan expected that experienced erection engineers, like Freeman or Minshall, might have seen a problem, but that he had considered the falsework "rugged."

Following Jestley, Farris began to question Swan about his company's contract with the Toll Authority, specifically page 19 of the specifications document where it mentioned falsework.

"It was never in any contract before . . ." Swan said, "but we have never done any job but what we followed through on the falsework, or we might—we reasonably assured ourselves of its adequacy."[174] It was only in the contract, in fact, as a residual clause from a previous contractor who had had two partial failures, Swan recalled, before speculating that had the falsework plans been submitted to his company, it is likely the mistakes would have been caught and the accident avoided.

Just before the Colonel was excused, Minshall requested permission to ask one question:

"Colonel Swan, you have in your evidence been very complimentary to the ability of Murray McDonald as a qualified erection engineer. Can you imagine him about one hour prior to the collapse taking only a casual look at the most vulnerable point in his erection scheme?"

"I found him very thorough when he worked for us," Swan replied. "[He] was always on the job. Any time that I visited the superstructure undertaken by Dominion Bridge Company he was there. I don't ever remember him being absent."[175]

When Frederick Dembiski mounted the stand next, he related that he was a special projects engineer for the Department of Highways under Fred Brown, the chief engineer, and that although he had been assigned the bridge project, neither he nor the Toll Authority had any responsibility for checking the erection or the erection methods.

"Did anyone else to your knowledge in the Department of Highways acting for the Toll Authority have any such duties?" Stewart, counsel to the Toll Authority, asked.

"No one," Dembiski replied emphatically.

Dr. Masters, who followed, was asked by Farris about practices in the United States. After establishing that he could only comment on the practices of his own firm, he suggested first that the AASHO specifications under which the bridge was being built was only meant to cover single-span structures 90 to 120 metres in length:

> It is on account of that limitation, I suppose, that we as consulting engineers are in business, because we deal with major structures that require the knowledge and skill and experience in the design of structures that exceed those that are commonly specified . . . We require then that the

contractor immediately upon being awarded the contract—practically all contracts are done on a competition basis—must submit to us his proposed method of constructing the work, and his plans, so that we can check it thoroughly and make sure that he has the capacity and the ability to do the work of the quality required ... we also require the contractor after we have approved the award of the contract to this man, to submit to us further details of his proposed methods of erecting the work, and he will then send us general plans and the general programme ... Then when he has satisfied us with his general scheme of erection, that it will meet all of these requirements, then ask him to further submit details of his erection plans, and those details we check and examine ... We do not approve. We don't propose to take the contractor's responsibility in any sense ... but we do take the right to examine these plans, and our examination is generally welcomed by the contractor.[176]

Masters's colleague J.R. Giese added to his testimony:

We request the contractor to furnish the plans for his falsework and his temporary work to our resident engineer. Our resident engineer is instructed to report any variation whatsoever in the construction of that falsework which may occur in the field, and to call the attention of our office to it so that some steps may be taken to ascertain why such a change has been made.[177]

Ralph Freeman, who took the stand after Giese, went one step further than his American colleagues. In his English firm it was standard practice, where falseworks were concerned, to insert a clause in the specification requesting the contractor to submit the particulars of the erection scheme. Then, once the contract was awarded, detailed schemes of erection were required including "working details, design details and working details of the various parts of the temporary structure, falsework, and so on that will be required." This also extended to the contractor submitting "properly written-up calculation sheets for those temporary designs to us for checking."[178] The sheets were stamped "Approved" if they were correct, but if they were unsatisfactory, they were returned with a letter of explanation of what was wrong. The second English engineer, Otter, discussed similar practices, though his firm didn't require calculations. Following his testimony, the commission adjourned for the weekend.

The following morning, predictably, newspaper headlines trumpeted the news: **BRIDGE FIRM ADMITS ERROR IN CALCULATIONS: DEAD ENGINEERS ARE NAMED**,[179] and **MYSTERY MAN SPOTTED TRAGIC BRIDGE ERROR**.[180] The *Victoria Times* reported that all efforts by the bridge company to find out the identity of the mystery man had failed.

John McKibbin's father Sam, still in Australia, after hearing the news of his son's name being made public, picked up pen and paper and wrote to John's widow Barbara:

> Our hearts go out to you and Mary McDonald at the news which came yesterday concerning the "findings" of the Royal Commission. We do not know enough detail to speak much. We did know, however, that it was a mortal—or rather an immortal—certainty that the buck would be passed on to the engineers. Economics and the needs of the living are paramount in such things. It is easy to blame the dead. They make no reply nor can they defend themselves. For our part our heads are held high in pride for our son and for our beloved and our hearts are at peace for him. Many factors could have entered into that calamity, such as faulty steel or welding; the tides; carelessness of some unknown workmen, and a score of other factors. The least likely would be the fault of two such careful engineers who prized human life above everything else.
>
> A few months ago John came to me in a dream. It was so natural. I said, "Who was to blame, John?" With that queer little shrug of his shoulders and a smile, he said, "Nobody knows, Dad." It certainly seemed to me in that dream—if dreams are messengers, and they are—that the last two who could possibly be blamed were Murray and John. But somebody must take the rap. I suppose a reason must be given in some way which clears the company. Some of John's statements in his letters home spoke of the carelessness of life in bridge building in Canada and reflected his disquiet about it. One statement he made was, "A man is worth three dollars an hour here and no more. Life does not seem to count very much; if any man is killed through any fault of mine I'll walk off the job."[181]

Although Eadie was the DB representative who had admitted his company erred, he appeared to be out of sync with his Vancouver colleagues who were still holding firm to their belief that the problem emanated firstly at the top of

Pier 14—even though they had agreed to the joint cause of the collapse at the beginning of stage three. By Monday, their thinking—in their minds anyway—would be vindicated by what they thought they had discovered after salvaging the falsework. They would also attempt to make their discovery very public. DB officials claimed to have found exactly what the prescient Freeman had suggested in his earlier, "secret" letter to Lett: sabotage. Far from being a stalling tactic, the company fervently believed that it had found the real reason for the collapse.

When the assistant royal commission counsel, Wallace, arrived at his office early Monday morning, he received a telephone call from a *Vancouver Sun* reporter wanting to confirm a press release that had just been issued by Mr. Stannard, DB's public relations officer. DB had discovered evidence of explosives being used on the east leg of the bent and the reporter now wanted a comment from the commission.

After DB officials had discussed the find with Sanderson, the company had cut out a supposedly concussed piece of steel and forwarded it to Montreal for analysis. Wallace phoned Jestley, who was equally in the dark, but later relayed to Lett that he had spoken with Gentiles and it was true. Lett recorded this conversation in his diary:

> I intimated to Mr. Jestley that I was very surprised that Dominion Bridge would authorize such a press statement whilst the Commission was still sitting and when it was about to conclude the evidence, and further suggested that if Dominion Bridge had any such information they should bring it before the Commission formally, and not make a statement to the newspapers.[182]

DB managed to call off the reporter, thus avoiding a public drubbing, but the alleged discovery caused the commission to pause. The legs would have to be re-examined by experts. Lett was now in a quandary. He was in the final stages of preparing his summation, but now had this disturbing new information. He convened a meeting with Wallace, Giese, Freeman and Otter and it was suggested that they take a look at the salvaged legs. Lett also contacted Brigadier Bishop at the Chilliwack army base and eventually secured the services of an explosives expert, Major Dunbar (retired). He would be accompanied by another military expert, Flight Sgt. Longstaff of the RCAF base at Sea Island. They would have until October 16 to inspect the bent and prepare their report.

Although Lett was planning to adjourn the commission for one day until

Wednesday, October 8, to accommodate Charles Andrew, a chief consulting engineer with the Washington State Toll Authority who had been too ill to appear earlier, the rest of the Monday was devoted to examining Cedric Saunders, a Department of Highways' bridge engineer, plus various other experts. Saunders, who mirrored Dembiski's testimony, advised that at no time did the provincial Highways department supervise or inspect the construction of the Second Narrows Bridge, nor make contact with the contractors—that responsibility was left entirely to the consulting engineers.

Herbert Barrett, a consulting engineer in the field of structural engineering, then took the stand. He advised that when his company prepared specifications for contractors, it was general practice for them to furnish plans for falsework, and jacking. He recalled that falsework plans had been requested for the Granville Street Bridge's concrete approach spans, and for the Oak Street and Middle Arm Bridges, but he could only recall seeing the plans for the Granville Street Bridge. When asked what responsibility a consulting engineer had with respect to falseworks, Barrett replied with, "It is a matter of professional responsibility to satisfy himself that nothing is being done which would endanger the structure or the people engaged on it."[183] However, he also offered the caveat that the onus was always on the contractor.

When William McKenzie, a contemporary of Barrett, took the stand next, Jestley asked him whether he knew Murray McDonald and what he thought of him. Wallace bounced to his feet to protest that it was not a proper question. The commissioner quickly responded:

> Well, questions have been asked about Mr. Murray McDonald's capabilities, and the impression I have from the evidence to date is that he was considered a most capable and most promising, I think were the terms Major Swan used, and that he was most thorough and most assiduous. Various terms have been used about him, and that is the impression I have in my mind of Murray McDonald at this point.[184]

In fact, McDonald's reputation was so strong that he was being sought after by other firms. Only eleven days before the collapse he received a letter from a Montreal executive placement service stating that "It has always been our policy to keep a close ear to the ground in our search for good men. In the course of this process, somebody mentioned your name."[185]

Later that afternoon the commission was informed that Lieutenant Colonel

Harry H. Minshall was withdrawing his services, after advising that he had nothing further to add to his earlier report. Perhaps he knew that it would be an uphill battle to convince the commission experts that the collapse had begun in the soft blocking assembly of Pier 14.

On that note, the commission adjourned until Wednesday when Charles Andrew would discuss his experience with respect to falsework plans in the State of Washington. Andrew was the engineer who had redesigned the Tacoma Narrows Bridge after the failure of the first bridge, which was only four months old when it collapsed before a modest wind on November 7, 1940. The former bridge was known fondly as "Galloping Gertie" because of its undulating behaviour, where motorists frequently reported that cars ahead of them would disappear entirely from view. The mile-long suspension bridge famously flew apart due to what engineers now consider to be an "aerodynamically induced condition of self-excitation."[186] It took them almost fifty years to arrive at that conclusion, abandoning the original cause of "forced resonance," after careful but deliberate study. Regardless, a film recording the event has educated thousands of young engineers around the world for almost seventy years.

Andrew stated that the Washington State Toll Authority followed the AASHO, and that their contract specifications informed a contractor that falsework plans may be required, but that the consultant would take no responsibility for them. He suggested that falsework plans would be requested from inexperienced contractors, but the opposite would probably be true for a large and experienced firm like DB. Not taking responsibility for falseworks, he advised, was a matter of "dual control . . . If we took responsibility we would have to check every little detail as it was done and that would take the job away from the contractor entirely."[187]

Following Andrew's examination, the commission rested until October 16 when details of the alleged concussed bent would be revealed. Meanwhile, across the inlet the scene was anything but restful. Traffic jams on the North Shore continued unabated. Even the RCMP were now offering opinions that the delays could only be solved by another bridge, an opinion that was shared by the British Columbia Automobile Association who were calling on the government to ready plans immediately for another crossing. Traffic was daily backed up to McKay Street in North Vancouver and was stalled all the way up Taylor Way in West Vancouver. North Shore officials were snarling that if the North Vancouver Ferries were still operating, the congestion would be more manageable. The provincial government, however, remained firm in its denial of aid to the city-owned ferry system, maintaining only that traffic would be eased once the new bridge was open.

In an effort to ease congestion at the Lions Gate Bridge, two approach lanes were added to the existing six, one for West Vancouver and one for North Vancouver. The eight toll booths on the bridge could now handle a record sixty-four cars per minute, up from fifty-six, though that largess exceeded the bridge's "flow through" capacity. Although Vancouver had solved some of its congestion problems a year earlier with the establishment of one-way streets, an experiment that was proving successful, there was little more that could be done to ease the bottleneck on the North Shore. A commuter's only option was to sit in traffic and fume, the convenience of cellphones and the comfort of travel mugs and CD players being decades away.

On the morning of Thursday, October 16, the commission reconvened to hear evidence from the demolition experts. They were united in their claim that there was no indication of explosives being used on the bent. William Armstrong, UBC professor, led the charge and came quickly to the point: "The investigation and the tests carried out by the writer have produced no evidence of damage caused by the detonation of an explosive charge in the east leg of the temporary support."[188] This conclusion was quickly corroborated by Major Dunbar.

Ralph Freeman sealed the case by agreeing with those who had preceded him, before clarifying his position, at his request, on the use of plywood as "softeners." In his opinion, the plywood was a contributory factor to the collapse:

> If the upper tier had been fitted with stiffeners and effective diaphragms the plywood would not have interfered with the intended performance of the grillage. However, the arrangement of this grillage was such that the pressure on the plywood was about 1,340 pounds per square inch under the west leg of Bent N4 immediately before the collapse. Now this pressure was shown by the tests carried out by Professor Hrennikoff to be beyond the elastic range of the plywood material. If plywood were to be used, therefore, as a soft packing in this grillage, the grillage should, in our opinion, have been arranged so as to keep the pressure on the plywood well within the elastic range of the material. This would have avoided the creep in the plywood which we referred to in paragraph 427 of our joint report, and would have [mitigated] the undesirably high lateral bending stress in the flanges of the upper grillage beams mentioned in paragraph 429 of the joint report.[189]

Freeman was then asked about Minshall's "soft packing displacement" theory, and quickly drove a stake through its heart: "Although we have considered that theory we found no evidence to show that the wedges and packs were displaced prior to the collapse; and the observed mode of collapse of the bridge is not consistent with such a prior displacement of the wedges and packs."[190]

By late morning, following all the evidence being presented and all the witnesses and experts heard, Farris proposed that the commission begin its summations. He quickly restated the joint report's conclusion, and then launched into a detailed argument, supported by the experts' evidence, refuting Minshall's theory. Further in his report, he caused Jestley to sit up and take notice:

> Now coming then to the question of negligence on the part of the Dominion Bridge Company and its employees, it will be my submission that there is ample evidence upon which your Lordship may find that the negligence of the Dominion Bridge Company caused the collapse of the bridge. It will be my further submission to you that the negligence causing the collapse, on the evidence, should not be attributed to any individual or individuals.[191]

Farris then provided the commission with Thomas Beven's 1880 definition, in trite law, of "negligence," which he described as "a legal duty to exercise care," and "a variation in the exercise of the care necessary under the circumstances of the particular case," which essentially meant "was the necessary amount of care exercised in this case." After quoting the section of the contract dealing with falsework, he continued:

> Well, your Lordship will see there are two obligations in respect of the falsework. First, it has to be properly designed and substantially constructed; and secondly, the contractor was required to submit plans of the falsework he proposed to use to the consulting engineer to enable the engineer to satisfy himself that the falsework was properly designed.[192]

Farris then reviewed the critical design sheet where he identified the errors, expounding on the fact that some witnesses had referred to the errors as mathematical in nature:

> In my submission it should be that your Lordship will find that they are

not mathematical errors at all; they are engineering errors . . . they also have been referred to as human errors—to err is human; that is quite true. That does not mean simply because it is human to err that the error does not constitute negligence . . . Now these errors were made, and the system of checking the Dominion Bridge Company had was ineffective as far as correcting these errors are concerned . . .

Now the responsibility for designing this important grillage was imposed on a comparatively inexperienced engineer, a young man who had only been with that company for three or four months . . . He had only graduated in 1954 in Sidney [sic], and therefore had a very limited experience, but the responsibility for designing this important grillage was imposed upon him. Now in that respect the usual organization chart of the Dominion Bridge Company was not followed. If it had been followed I think it is safe to say that the falsework would have been either designed or checked by one of the following—by a constructional engineer, a design engineer, the erection engineer, or the field engineer.[193]

Even though the design sheet had been checked by the erection engineer, Murray McDonald, Farris related "it may be that on the evidence which your Lordship will consider that an imposed responsibility was placed upon Mr. McDonald, that he apparently was expected to exercise two [four] functions: that of structural engineer, design engineer and erection engineer; as well as that of field engineer."[194] DB's practice of removing major projects from the company's normal channels, although perhaps expedient and convenient, had sidestepped one of the most fundamental tenets of the engineering profession: the critical checking process.

It wasn't a great leap from there to Farris's conclusion with respect to the giant steel erector:

I therefore submit, my Lord, that it is open to your Lordship to find the collapse of the bridge is due to the negligence of the contractor, Dominion Bridge Company, and that that negligence consisted in:

Faulty design of the upper grillage at Bent N4 as a consequence of two engineering errors;

In leaving the design of this important falsework to a comparatively inexperienced engineer;

In failing to provide within the organization an adequate system of checks of the design; Finally:

In failing to carry out its contractual obligation to submit the design of the falsework to the consulting engineers.

I further submit that it is open to you to find that the negligence is a corporate negligence. I think I have heard the expression used that this is a team effort and that the negligence should be assessed against the company rather than against any one individual.[195]

Farris then moved on to the consulting engineers, Swan, Wooster & Partners, where he was not quite so confident in his findings. After establishing that had Swan's firm reviewed the falsework plans and calculations, the errors would have been discovered and the accident avoided, Farris then posed the question, "Were the consulting engineers negligent in failing to check the plans of the falsework?" Putting forward two arguments, both having merit, one for the consulting firm being negligent and one for it not, he left it up to the commissioner to make the final ruling on the matter:

> As to the question of whether or not it was faulty judgment on the part of the consulting engineers, I find difficulty in distinguishing between faulty judgment in this connection and negligence. If the wording "errors in judgment" had been used, it would have perhaps simplified the matter, but in my view, if there was negligence there was faulty judgment, or if there was faulty judgment there was negligence, and I think that the issue really comes down to was there a duty on the part of the consulting engineers to check the falsework. If there was, admittedly that was not complied with, and in my view there would be negligence. If there was no such duty, then there was no negligence.[196]

Just before the commission recessed for lunch, Herb Macaulay from the Vancouver–New Westminster District Building Trades Council, and Norm Eddison, submitted their report complete with recommendations, stating that the loss of lives would be in vain if no changes were made to construction practices:

> All plans, drawings, calculations, methods of operations, sequence of operations, etc. on all major and/or minor structures (where a basic hazard may exist) be submitted to and approved by the Consulting Engineers or Architect before any erection either on the main part or incidental part of the general erection or construction scheme is carried out.

That all materials designed, fabricated and specified for use in a permanent structure shall not be used (except in case of extreme emergency) for any other purpose than that for which it is intended on the plan of the structure—Materials for temporary structures for the purpose of assisting in the erection of permanent structures should be considered expendable.[197]

Eddison suggested further that WCB should hire its own consulting engineers to examine structures in the future, a position that WCB lawyer Ralph Sullivan strenuously objected to, stating "it was not the duty of the Workman's Compensation Board to relieve management of their responsibility."[198]

Jestley now took the floor to present his summation, but before beginning he advised the commissioner that he wanted to address a few of Farris's points, the first being the matter of the errors being engineering and not mathematical in nature. After stating that the errors were "clearly mathematical errors caused by human error," he supported his argument with the fact that the *lower* grillage design sheet had been reworked by McKibbin before being approved by McDonald. This was proof enough, he opined, that the engineers were competent, and had erred in their math only, not in their engineering when calculating the needs of the upper grillage. His second objection had to do with Farris's comment that DB had left the falsework design to a "comparatively inexperienced" engineer, to which he countered with McKibbin's adequate level of experience, and the previously stated fact that the relatively simple calculations could have been prepared by a fourth-year engineering student.

Arguing Farris's next point about "checking," Jestley recalled Eadie's description of McDonald and how competent and thorough he said he was, and that he had been specifically chosen for the job. And with respect to the company's failure to submit the falsework plans, Jestley contended that "it certainly has been brought out in evidence that the practice is to submit the general scheme of erection and only falsework particulars when they are demanded."[199]

At the end of his rebuttal, he pointed to Farris's description of DB's negligence as "corporate negligence," a term that he advised he had no knowledge of in law but admitted that he clearly understood what was meant by it, given that it was trite law:

A corporation, like any other employer, is under a duty to select agents or employees competent to discharge their duties. Once having done so, a corporation cannot then itself be guilty of negligence except through the negligent actions of its servants committed in the course of their employment.

Thus for the purpose of indemnifying third parties who have suffered damage, the negligence of the servant is attributed to the corporation. Unless the actions of these servants are in themselves negligent, however, there can be negligence on the part of a corporation which has exercised reasonable care in the selection of its employees for the work of which they are to be engaged.

My learned friend has pleaded with you that you should not find any negligence on behalf of the employees. I am glad to go along with him on that, but I think a statement of the law shows that if there is no negligence on behalf of the employees there is certainly no negligence on behalf of the company.[200]

Jestley then launched into the body of his report, suggesting that DB had been fully cooperative with the commission and that it had always been in agreement with the findings of the royal commission experts. This position was somewhat surprising given his efforts to manoeuvre attention away from the upper grillage as the primary source of the collapse, instead toward a release of stresses in the blocking assembly of Pier 14—extending even to the belief of sabotage. He suggested that support for his clients' acceptance of the facts was his decision not to cross-examine the experts on the joint report. He then tackled the question of human error, which he said all men are subject to:

Despite the errors, the two engineers were competent ... all of the men with whom they worked and the other engineers with whom they came into contact had only the highest praise for them and it is difficult to understand how these errors occurred, particularly, my lord, when you refer to the excellent work done on the lower grillage on the very same day that the work was done on the upper grillage.[201]

His conclusion was clear, if somewhat predictable:

In this case the collapse did not occur through lack of knowledge or know-how or lack of experience nor lack of efficient equipment nor lack of sound erection procedures; it occurred because of human error. As I have stated, it is comparatively easy to point to the cause after the event. In the present case the examination of the brief sheet dealing with Bent N4 immediately pointed out the cause. This accident proves that no matter how competent the organization or the extent of the precautions taken even the best-run organizations and the most highly qualified engineers are subject to human error. This is certainly a risk which must be assumed at all times and I submit it is a risk inherent in the engineering profession, my lord.[202]

No matter how eloquent Jestley's argument was, the press didn't buy it. They only had ears for Farris when he attributed the collapse to the negligence of the giant steel erector. A special edition of the *Vancouver Sun* newspaper, issued later that day, heralded the recommendation in bold capitals: **BRIDGE FIRM ACCUSED OF NEGLIGENCE**. DB's response was to repeat that it was "human error" and that "In view of the fabrication methods, erection procedure, competent engineers and precautions taken by the company, nothing more should have been reasonably expected of it."[203] It couldn't get much worse for the bridge company, but Farris's ambivalence toward finding the same for the consulting engineers saved them from a similar fate . . . for now.

Steer, counsel for Swan-Wooster, when taking the floor from Jestley, dealt first with the question of liability between the consulting firm and the Toll Authority, which he quashed with the comment from case law that "the contractor, and he alone, is responsible notwithstanding the fact that plans have been furnished to him." Steer then moved on to the question posed by Farris of "whether or not the obligation of the engineers to the Authority which employed them has been satisfied,"[204] or in other words, were the consulting engineers negligent in failing to check the falsework plans?

Acknowledging that the contract had two obligations—"plans, specifications, tenders" and "supervision"—Steer admitted that there was no question about the first part, but with respect to the second there was a question about breach of duty and whether it was negligence or faulty judgment. Given that the contract lumped falsework in with materials, tools and labour, which were considered incidental to the erection, he argued that his client was not responsible for the details of any of these items. Other sections of the contract dealt with the main erection and working details, but had no request for specifics. If an erection procedure drawing had been ready and had been provided to the consulting engineers as was the practice, it would not have shown much anyway, Steer argued, except a drawing of the legs of the bent—hardly enough information to draw any conclusion from:

Now here is the situation: the engineer is, under this specification, to get an erection procedure drawing. He does not get it. What he does get is a verbal statement which will give him all the information, in my submission, which an erection procedure drawing would give him. The engineer had checked the foundation for the bent. He had been informed that a bent was to be used. From twenty-five years experience with the Dominion Bridge Company he had confidence in their competence to do a piece of work which not only he but other experts who gave evidence to your Lordship regarded as a routine piece of work. One cannot help remarking here that no engineer could contemplate the errors being made in the office of the Dominion Bridge Company in the manner in which my learned friend Mr. Jestley reported these mistakes to have been made just a little while ago.[205]

Steer continued, quoting case law and the royal commission transcript as further support of his argument, emphasizing Andrew's comment of dual responsibility, of not relieving the contractor of their responsibility, and finished with his belief "that the duty was fully performed and that there is no responsibility on them for this unfortunate occurrence, legal, moral or other."[206]

Just as he was concluding his case, Farris was handed a note from Norm Eddison, who was now representing Local 97 and wished to give another statement. As no objections were made, Eddison began:

I further submit that a detailed drawing and plan, a blueprint, of this falsework by necessity must have been in existence prior to the erection of N-4 by reason it is impossible to build such a thing without a plan, and with that, your Lordship, I think I will conclude my remarks. There is one more point, possibly, that it was suggested by a witness some time ago that these plans were not ready. It would not seem logical to me that plans would not be ready some months after the structure had been erected and, the final point would be that it would seem that the Dominion Bridge Company were remiss in their duties and requirements of contract in not submitting such plans of falsework, secondly, that the consulting engineers, Swan, Wooster & Partners, were remiss in their duties in not asking for plans of such an important part of the structure.[207]

Hill, counsel to Swan-Wooster, responded immediately, asking that the commissioner ignore Eddison in favour of Minshall, as he spoke with "more authority." Eddison promptly replied that he and Minshall were representing the Ironworkers Union together.

Before Lett adjourned the royal commission, Farris asked to speak for five minutes to refute Jestley's arguments against finding DB negligent:

> My friend has submitted that there was no such thing as corporate negligence, and he cited the law which I think we all know that a company only acts through its servants and agents. My submission to you, my Lord, was that the negligence could not be attributed to any one individual and I say on the evidence you might find it difficult to pinpoint the negligence and to say who it was that appointed an inexperienced, comparatively inexperienced man to design this important falsework, who it was that imposed what might be considered perhaps impossible burdens on Mr. McDonald, who it was that was responsible for failing to provide an adequate system of checks, and who was responsible for the failure to submit the design of falsework to the consulting engineers, and I simply suggest that your obligation under the terms of the Commission to determine whether or not there was negligence or faulty judgment on the part of any person is satisfied if you find that there was negligence on the part of the Dominion Bridge Company.[208]

The last words of the regular session would be uttered by the commissioner himself. Thanking all who had participated, he declared that although there were discrepancies in the eyewitness accounts, he believed that they all were honest. He thanked DB, Swan-Wooster, Kiewit-Kennedy (Kiewit-Raymond while building substructure), Local 97, the Painters Union, the Toll Authority and the various experts for their assistance, and expressed his gratitude to DB and Swan-Wooster for coming forward before the commission began stage three to inform him that they were both in agreement with the experts' findings. He saved his last thanks for all those who had participated in the rescue, and then his commission staff, before adjourning the royal commission to finish preparing his summation.

While reworking his report, a question came to mind: What would the practice be with respect to falsework plans for the balance of the erection process? Swan advised that he wouldn't be checking them unless the Toll Authority instructed him to. Lett believed that consulting engineers should be requesting plans when they weren't freely offered by a contractor, and felt that it was his duty to make a specific recommendation to that effect. Lett summoned Jestley and Farris to his office to determine what DB planned to do, and was informed by Jestley that DB fully intended to submit falsework plans to the consulting engineer, and would not be embarrassed by such a recommendation. After Lett read Jestley the draft of his proposed findings, he recorded Jestley's reaction: "while there was a breach of a contractual obligation to submit the plans, he did not consider that that was negligence but that he could not object to my finding that the Bridge Company was responsible for the collapse."[209] How could DB be negligent, Jestley argued, given that the large steel erector had built over three thousand bridges across the nation without a single collapse except for the Duplessis Bridge, which he said was due to a bomb.[210] Although impressive, it was not enough to sway the commissioner.

On December 3, Lett reconvened the royal commission to present his report. After restating the already published cause of the collapse, including the contributory factor of the plywood softeners, the commissioner found that the Toll Authority had neither received nor given any information with respect to the erection and that "There was no negligence or faulty judgment on the part of the Authority which caused or contributed to the collapse." The Toll Authority, and by association the government, were now publicly absolved from all interference or blame.

Lett then stated that the commission had used the term "negligence of any corporation," and that there was some question whether "negligence in the legal sense can be found against an incorporated company, as distinct from the negligence of its servants or agents." He decided that it was unnecessary to make such a distinction given that "negligence of its servants and agents is a matter for which a corporation may, under certain circumstances, be found responsible." Lett added that the collapse was indeed caused by negligence and that "It remains then to determine what was the negligence and to fix responsibility for it."[211]

He then began a summary review, highlighting facts like the terms of the contract which contained two obligations: to properly design and construct the falsework, and to submit the plans to the consulting engineers. He commented on Eadie's testimony regarding DB isolating major projects. He then talked about the two engineers in complimentary terms before highlighting details of the critical design sheet. "In the absence of explanations from the engineers who perished in the collapse, it would appear that the erection of the grillage without stiffeners and effective diaphragming originated in the error in the calculations shown by the calculation sheet," he said. "But, in my view, it would be unfair to those who had to do with those calculations, and an oversimplification, to say that it was only a mathematical error which caused or contributed to the collapse."[212]

Lett returned to DB's practice of isolating major projects, and how that left a comparatively inexperienced engineer to design the falsework with only one person checking. This, Lett said, absolved the company of its responsibilities, "and the ordinary provisions for checking falsework as provided in the management guide, were not in force."[213] He then found the erection company negligent for: a) failure to properly design and erect the falsework as per the contract specifications; b) failure to submit the plans to the consulting engineers; and c) leaving the design of the falsework to a comparatively inexperienced engineer, with inadequate checking.

The commissioner then dealt with Swan-Wooster, noting that the contract specifications, drawn by the consulting engineers themselves, called for the falsework plans to be submitted to them. He related Wooster's testimony where he advised the commission that the consulting engineers tested bolts, checked steel members, shop drawings and painting, but did not check the falsework. He noted that Swan took little notice of the fact that he wasn't provided with the N4 and N5 falsework plans, given that the onus was always on the contractor, and repeated Swan's disclosure that if his firm had seen the plans the accident likely would have been avoided. "In view of their long and successful association with the Contractor, this failure is understandable, but it cannot be overlooked. Accordingly, I must find there was a lack of care on the part of the Engineers in not requiring the Contractor to submit plans of the falsework."[214] Although a "lack of care" was not as damning as the crippling word, "negligence," it would burden the consulting firm as though it were. Swan would never build another bridge.

The commissioner's recommendations followed:

> In the light of what has happened, I recommend to the Authority that it requires the Consulting Engineers to demand of the Contractors all their proposed plans (including computations and stress analyses showing the strengthening of the structure itself in order to provide for safe cantilever erection without falsework, and any devices used for effecting the closure of the cantilevered span) for the erection of the remainder of the work to enable the Engineers to examine those procedures and details. The Engineer's examination should include, by way of precaution, an examination of the functioning of the tie-down devices in the well of Pier 14. The Contractor should not use any temporary works in erection

until the Engineers have signified that they have completed their examination and have no objection to the Contractor's proposals.[215]

The report, once made public, was electric. It triggered newspapers across the nation to repeat what Farris had already suggested would be the finding of the commission. Both firms argued the claims; Jestley, for DB, suggesting that nothing more could have been expected of the firm, and that the errors were "human" and not "engineering deficiencies," but "a risk inherent in the engineering profession," and Steer, for Swan-Wooster, repeating that it was not the consulting engineers' responsibility.

Premier Bennett reacted strongly to Lett's findings, attacking Donald Cromie, publisher of the *Vancouver Sun* newspaper, and Opposition leader Robert Strachan for their criticism toward the Toll Authority and the Minister of Highways, Phil Gaglardi. Suggesting that both men should resign, he stated that the *Sun*, in a June 18 editorial, "tried to create public uneasiness" on other bridges, and had "smeared Mr. Gaglardi as minister of highways," and as well, Mr. Strachan "tried to throw suspicion and smear on Mr. Gaglardi."[216]

"Never has the government moved so quickly to set up an enquiry into a disaster,"[217] Bennett said triumphantly of the $50,000 royal commission, as he revelled in the fact that the Toll Authority had been given a "clean bill of health." Gaglardi then affirmed that DB would be responsible for the $3 million cost of repairing the bridge which he expected would be finished about four months behind schedule.

Although Lett had now concluded formal commission business, he was perhaps not quite excused from duty, as he had the sad task of having to listen to an angry telephone call from Colonel Swan's wife, recorded in his diary on December 3. Telling Lett that he was "crucifying my husband" and that it was "a grave injustice to Bill," she declared that he was "quite right in finding against [DB]" but that "you showed no mercy to my husband." Threatening that she "had some other things and I am going to bring it to light," she advised him that she was seeking legal counsel. Time, and perhaps the intervention of her still-respected husband (who had been unaware of her call), calmed her fury, which was perhaps more trying on the commissioner given that Bill Swan was an old friend.

11

Strike

Sometimes you just have to steel your nerves to get the job done. Don Jamieson, a DB engineer, felt that way as the dive tenders placed the "John Brown" rig over his head to rest on his broad shoulders. Although this would be a scary proposition for any initiate, for Jamieson it was doubly troubling ever since he had been invalided, temporarily deaf and blind, off HMCS *Saskatchewan* as a young petty officer during World War II. That episode triggered a form of claustrophobia in him that being entombed in a brass helmet didn't much help. But Jamieson, who had just replaced his friend Murray McDonald on the bridge as field engineer, was determined to see for himself what the tangled mass of steel looked like below the waterline. Jamieson and McDonald had worked together on the Lions Gate Bridge over twenty years earlier and the two had become fast friends. The thought of McDonald's death, as well as that of his old university pal, William Mck. Swan, on the Pattullo Bridge in 1937, was unnerving, but that didn't stop him from doing what he had to do.

It was obvious from the start that bridge building would become Jamieson's life's work, his father being a proud and vocal ironworker. Ditto that for his father-in-law, Bryce "Scotty" Philip, whose mantra that "ironworking is the only real man's job," would fail to sway the eager student from his goal of becoming an engineer. Although Lieutenant Colonel Harry H. Minshall, DB engineer, had convinced Jamieson to study civil engineering, it was the affable chief engineer, Angus McLachlan, who would become his mentor, who instilled in him a love of bridge engineering.

Offering to dive on the wreckage was also perhaps a manifestation of some kind of misplaced guilt in Jamieson. He was only too aware that it might have

A tilted Pier 14 stands aloof in front of the almost completely dismantled Span 4, the now calm waters of the Inlet hardly reflective of the tragedy that occurred here months earlier. OTTO LANDAUER OF LEONARD FRANK PHOTOS, JEWISH MUSEUM ARCHIVES LF-35951

been him who died on the bridge, given that he had switched jobs with McDonald a while before the collapse so that he could spend more time at home with his wife Jean and four young children. Jamieson had already spent thirteen years in the field and it was time for him to come home. Field engineers were known to clock a few miles. Now that DB had given Jamieson his marching orders, he was damn well going to live up to his obligations.

Perhaps some of Jamieson's anxiety also had to do with the recent loss of Len Mott, whose body was lost and presumed drowned. At this time, Mott's wife was simultaneously appearing before Justice A.M. Manson to have her husband declared deceased so that she could get on with her life. Justice Manson could see no reason why Mott's widow could not proceed with settling his affairs given that he had been missing for almost four months. Although the judge's decision would satisfy Mott's widow, within half a year Manson's name would surface with less respect on the tongues of the ironworkers who were embroiled in a bitter battle against a draconian piece of labour legislation that the judge appeared delighted to enforce.

Traveller No. 3, mounted on a small wooden barge, assisted with the dismantling of Span 5. OTTO LANDAUER OF LEONARD FRANK PHOTOS, JEWISH MUSEUM ARCHIVES LF-35952

As he was lowered beneath the surface, Jamieson fiddled with the buoyancy valve on his dry suit to find some stability, experiencing a moment of panic when he voided all the air in the suit, causing water to instantly jet into his helmet. Staring nervously through the small barred porthole, he "realized what a slender thread that connects these men with life."[219] Although visibility was only a scant few feet, the twisted steel loomed out of the green darkness like a wrecked ship, offering the gutsy engineer with a fresh perspective on how to dismantle it safely.

As Jamieson was later helped from the water by the dive tenders, he must have been mulling over the fact that dismantling the twisted spans would not be quite as easy as anticipated. For one thing, the stresses on the steel had been reversed with the collapse. Where there had been tension, now there was compression, and vice versa. Knowing this made it easier, but then the buckles and bends in some of the members added new dimensions to the stresses and the risk. In order to normalize the stresses, a steel A-frame designed by Bob Harris, who was now the erection engineer, was attached to the bottom of the wrecked spans. Jack brackets were also installed to accommodate the six 287.8-tonne hydraulic jacks that the A-frame, or secondary truss, would take reaction from. When the frame was pulled, or jacked opposite to the stresses, it reversed them back to what they had been before the collapse, permitting the ironworkers to dismantle the twisted structure with a greater, but not always reliable, degree of certainty.

Starting at the top of the tipped spans, each piece was unbolted and lifted by Traveller No. 3 down onto an adjacent wooden barge. Another crane, an antique with a big wooden A-frame known as the Arrow Giant, was mounted on another barge to facilitate lifting smaller members. It was steam driven and whenever it made a big lift, the end of the barge would become awash. The operator, who kept several crab traps near the site, would use one of the steam lines to cook his catch at lunch. He was a popular lunch partner.

Once salvaged, the pieces were field inspected, numbered, and if they were undamaged, deposited in a yard just off the north bridge approach where Bob Dolphin, one of Jamieson's two assistant erection engineers (George Shephard being the other), organized and kept track of them for later recovery. Dolphin recalls that "part of the problem was making sure you could find everything . . . a lot of little things . . . little clip angles, things like that." If they were at all damaged, but salvageable, they were loaded aboard a railcar or truck depending upon their size, and shipped to the plant to be refabricated before being returned. Each piece, before it could be reassigned to the bridge, was then personally inspected, metallurgically tested, and endorsed by Allan Kay. The pieces that were too badly damaged were relegated to the scrapyard.

The north-side steel close to the same position it was before the collapse. OTTO LANDAUER OF LEONARD FRANK PHOTOS, JEWISH MUSEUM ARCHIVES LF-36692

Dismantling the top part of the downed spans was easy, but as Don Heron recalls, when they got down to the twisted steel at the bottom of both spans, it was a different story:

> You went down with a torch, a burning torch, and you had your goggles and that, and you went down on the ball of the crane where the iron was all mangled up and you set a choker on it, on a piece where you figured it would be balanced, and then you just kind of swung around and walked around and cut, cut a chunk of it out and then it was lifted out. And then, you know, you'd ride it up, or you'd find a spot where you could stay off and they could pull it out, but that was kind of awkward because the iron was all twisted up . . . a lot of it would be under tension and you never knew whether a piece would spring up at you or spring away, or drop.

Working alongside Heron were several of the collapse survivors: Bill Stroud,

Norm Atkinson, Charlie Geisser, Colin Glendinning and Bill Wright. Lou Lessard was there too, but as he was still on crutches, he was restricted from taking a physical role. Jim English was back on the job as well, his habit being to walk the bridge early. When the men arrived he could be found sitting nonchalantly in the foreman's shack with his feet on the table reading the morning paper. "He was as cool as a cucumber and had complete command of the job," Bob Dolphin recalled.

There was no thought of these men not returning to work. Glendinning said it best when he commented to a *Vancouver Province* reporter on the eve of the first anniversary, that "You have to climb back on the horse that threw you to show who's boss."[220] But for some of the other survivors, their careers had been nipped in the bud by the collapse. Gordy Schmidt would never work in the trade again on one leg, one ironworker couldn't control a newfound fear while another's wife convinced him to find a safer occupation.

Survivors one year later. From left to right, front row: Bill Moore; Bill Stroud. Back row: Bill Wright, Jim English, Lou Lessard, Charlie Geisser, Norm Atkinson, Colin Glendinning. PHOTO P. STANNARD, DOMINION BRIDGE CO. LTD.

Once the ironworkers had lifted the last of the twisted sections from the inlet, Don Jamieson and his two assistants began the detailed work of planning the reconstruction. It had been made very clear to Jamieson that he, Dolphin and Shephard would be responsible for erection only, and given their general state of activity, that is all they could have handled. While Murray McDonald had been the field engineer, erection engineer and design engineer, Jamieson and his crew had only one job—erection—the rest being handled by other DB engineers domiciled at the plant. DB had obviously been attentive during the royal commission.

Jamieson's first order of business was to measure the distance between Piers 15 and 16, upon which would rest the long clear cantilevered truss comprising the north and south arms plus the suspended "drop-in" span. The distance had never been measured before, the pier locations having been set by triangulation. Jamieson wanted to determine the exact distance before steel erection would begin. There was only one way to measure it and that was to string a calibrated piano wire across the inlet between the two piers. Dolphin remembers that "the layout part was interesting because we did it at Hooker Chemicals there, alongside their fence, and we had to support the chain every ten feet I think it was . . . I mean, it took a month and a half just to do that." Using a standard 30.5-metre shop chain over the required 335-metre distance, standard tension had to be applied to the chain as well as an adjustment made for standard temperature (21.1 degrees Celsius). On a cold day the chain would be short. Dolphin recalled that Jamieson was meticulous in his measurements, perhaps too much so; he "made bloody sure everything was checked. That's all we did was check, check, check, check . . ."

Then, early one overcast Sunday—overcast so that the sun's heat wouldn't expand the wire—when marine traffic was at its minimum, two large tripods were erected to support the wire. 23-kilogram weights had to be applied to each end to minimize sag, and the breach was accurately measured. Dolphin recalled "that the actual measurement of the two piers was within three-sixteenths of an inch of the triangulated setting."

While Jamieson and his crew were busy with their planning, the families of the lost men were only just beginning to feel the inevitable financial pinch that came with their small WCB pensions. Gordon MacLean's widow received $100 per month but had to pay out $85 a month in rent. "I just can't go on like this much longer," she said. "I have been trying to get a job but it just seems none is available." Other wives had similar disappointment in locating work. But Mrs. Chrusch, who had four children to feed and clothe, didn't let her measly income get in the way of her ambitions. Purchasing a house with her share of the Families

Fund, after her landlady insisted on her paying the rent two months in advance, she believed that being cheerful healed a multitude of pain. "It's no use feeling sorry for yourself, not when you have children to take care of," she said.[221]

As the year drew to a close, the *Vancouver Sun* listed the "Year's Big Ten," the most influential stories of the year. The Second Narrows Bridge disaster was number one, followed by other events such as the annihilation of the infamous Ripple Rock and an exposé on Doukhobor violence. The bridge story continued to dominate the news the following year as the media and the city watched with interest as the two downed spans were dismantled and the site cleared for the inevitable rush to erect the steel, now four months behind schedule. Construction on the south side was expected to start on March 2 with an anticipated mid-August completion date. But this wasn't fast enough for Fred Brown, the Toll Authority's chief engineer, who sent a stern letter to Wooster telling him to warn DB that "I do not need to review the Toll Authority's position in this matter other than to say that the loss of revenue with which it is faced is of utmost importance to their operation." The letter complained that pier reconstruction had yet to commence and that the government's own inspection had indicated "an unreasonable delay in the erection of steel from the south abutment."[222]

Although the company was working as fast as it could, it had been dealt a staggering blow by the collapse, and was determined not to make another mistake. Erection of the south-side steel, once started, would be easy; it was the dismantling of the wrecked steel and repair of the piers on the north side that was causing delays. Even then, the finished structure was only expected to be five months behind its original completion date.

A few days after the start of steel construction on the south side, demolition experts Northern Construction Co. and J.W. Stewart Ltd., which had been subcontracted by DB, began to demolish the twin columns of Pier 14. Pier 13, also damaged, was repaired with strategic welding and concrete patches. Small holes were first drilled at the top of the Pier 14 columns to receive the charges that would slowly blast away its height. The first detonation brought down the heavy crossbar. As each explosion exposed the steel reinforcing iron, ironworkers burnt it away with acetylene torches. In five-foot sections, the total demolition was expected to take about three weeks. When the columns were finally down, the engineers decided to demolish the pier's pedestal as well even though it showed no visible signs of damage. It rested on a much broader, steel-jacketed concrete base, which would remain to support the new pedestal and columns.

Walter Mykietyn, project superintendent for the two construction/demolition firms, set off the 545 kilograms of Forcite he had packed into a honeycomb of 7.6-metre holes drilled into the pedestal. Only a third of it ignited, demolishing the western side of the concrete support plus a section of the adjacent wooden trestle. As timbers from the trestle twirled down into the inlet, he mused that he "didn't expect so much damage."[223] The second try demolished the other end of the pedestal and part of the middle, but still left the veteran demolition expert perplexed as to why it was being so stubborn. He decided, however, that he could probably remove the rest of the concrete without the necessity of further explosives.

Across the inlet, as the ironworkers working on the south side looked up occasionally to watch and listen as the pier was gradually being blasted away, this interesting distraction did little to divert their attention from a far more serious concern. They had been without a contract since the previous December, and were now rejecting a conciliation board's recommended forty-four-cent increase—over twenty-seven months—on top of their average $2.62-per-hour wage, saying that "The union wants an increase 'commensurate with a most hazardous occupation.'"[224] DB was now worried that the union would picket the bridge before it was finished, and in anticipation of this, together with Bickerton Bridge and Steel Erectors and Pacific Steel Erectors, three of the largest steel erectors in the province, approached the government to apply for a government-supervised strike vote. The union was in simultaneous talks with about twenty firms at the time, and the men were biding their time for an advantage. The bridge was shaping up to be their most visible and powerful leveraging option.

Less than a month later, close to the eve of the first anniversary of the collapse, the steel on the north shore had advanced no farther than Pier 13, and perhaps, given the current labour climate, it would remain that way for some time. Forty members of Local 97 working on the bridge were about to take a strike vote, action that would likely be followed by as many as 350 of their brothers working on other jobs throughout the province. Dominion Bridge, expecting a protracted strike, as the two parties were far apart in their negotiations, made plans to secure the steelwork for an almost certain extended shutdown. But the Local first had to make sure that its members had the stomach for a strike.

In a government-supervised process, ballots were mailed to members working on the Peace River Bridge at Taylor, and to a hydro project at Whitehorse. Once they were returned, ballots would be issued to Lower Mainland workers. Then, if approved by the membership, a strike could be called with forty-eight hours' notice any time within the next three months. A strike would affect construction all over BC and the Yukon: the Peace River Bridge at Taylor, BC Electric's Bridge River project, the hydro project at Whitehorse, North Vancouver's

On strike! Dougal McDonald (left) and Al Legere beneath the south-side steel.
PHOTO BILL CUNNINGHAM, *THE PROVINCE*

Lions Gate Hospital, BCE's thermal plant at Ioco, as well as many other large and small projects. It would also affect hundreds of other unionized labourers who would be obliged to honour the ironworkers' picket lines.

When the votes were finally counted a week later, it was an overwhelming 82.6 percent in favour of a strike, ironworkers on the bridge alone voting a resounding seventy-five to four. To the men, it was a matter of principle, a matter of honour. Existing wages were hardly reflective of the danger they faced daily, thus their request for a sixty-cents-per-hour wage increase over their existing hourly rate, but a bigger issue was that DB was trying to replace their travel-time pay with a six-dollars-per-day living allowance, essentially rescinding 50 percent of the forty-four-cents hourly wage increase it was offering. Also, the company wanted the option to import up to 50 percent of its workforce from Alberta.

Local 97 business agent Norm Eddison, who was known as "ten spot" because he always had a ten-dollar bill in his pocket, angrily commented that "Our union has suffered terrible unemployment, and we just won't go for bringing men in from outside."[225] Although the union, now six hundred to seven hundred strong, had planned to continue to negotiate, a breakdown in talks, an intractable company and the strength of the strike vote motivated it to submit its strike notice to the Structural Steel Association on Friday, June 19. The first anniversary of the collapse only two days earlier was enough of a reminder of the danger the men faced daily, ample evidence for a wage that matched the risk.

The weather on that first anniversary, in contrast to the tragic day a year earlier, had dawned mostly cloudy with mainly clear skies in the afternoon. For most of the men working on the bridge, it was just another workday, though it did give some pause for thought. Tony Wohlfart, a Boshard painter who went down with the bridge, said it best: "a man has to be stupid not to feel some fear. Some people clean up streets, some close up cans and we work in high places. Nobody forces us to do this. There is always danger in bridge work—in many types of work." Laci Szokol responded in like form: "Afraid to go back to the job? No, not really. I like high places. The view is nice," he said matter-of-factly, grinning.[226]

Although the overwhelming strike vote had handed the Local a strong imperative, it was just as vital to the government who had only a few months earlier, at the start of the 1959 legislative session, passed a new Trade Union Act, Bill 43, which was meant to curtail labour action. The session was even nicknamed the "Labour Session." The government was now about to test it, though Gaglardi claimed the Province wouldn't interfere and hoped that the two parties could work out their disagreements amicably. His plan was likely just to sit back and let the new labour bill do what had been intended of it. At the same time, though, he cautioned in a classic understatement that a strike would cause a delay, noting that "There is a certain amount of congestion and a great need for the bridge."[227] No doubt he had witnessed the honking cars and angry voices of commuters locked in the daily traffic grind that frustrated much of the population of Vancouver and the North Shore.

It was no news that the labour climate in BC at the time was tense, and it was also no news that BC harboured more labour activists than any other province in the country. At the time, fishermen and tendermen were threatening to strike, as were shore workers and the National Association of Marine Engineers. These, added to the large and powerful voice of the International Woodworkers of America (IWA), which was already on strike, made for a tumultuous time. Perhaps what irked the premier more than the control he felt labour leaders had over their membership, was the fact that they had found a new pulpit and were quickly becoming politicized. The unions were aligning with the opposition CCF party, a party whose power was ironically growing stronger with the prosperity generated by the Socreds.[228] Bennett's government was called anti-labour, which Bill 43 did nothing to dispel. In fact, the new Trade Union Act was as narrow an endorsement of the working man as the previous bill, Bill 28, had been when it was passed with as much controversy in 1954. That bill caused such a furor in labour circles that it prompted a delegation of union leaders to descend on Victoria to protest it. The new bill was indeed restrictive, and subject to loose interpretation, which would soon be evident in court.

The government wanted it both ways. Bennett had nothing against unionized workers, in fact actively courted them as part of his election strategy, but he had no time for the organizers, the bosses. Unions were considered okay as long as they didn't interfere with the normal course of business, but as soon as they did, their demands were considered unrealistic.

Phil Gaglardi called labour leaders "agitators," stating that "We don't need any Hoffas or gangsterism in this province."[229] Bennett sold the bill to the public in the polite but diluted terms of diplomacy, telling them that it merely imposed new responsibilities on labour and management. This, of course, didn't wash with the ironworkers on the picket lines, fighting to better their lives against employers wanting to pad their bottom lines at the ironworkers' expense.

On Tuesday, June 23, the ironworkers struck, shutting down dozens of projects across the province with a combined value of over $90 million. Out of respect for the emergency care the collapse victims had received at the North Shore General Hospital almost exactly a year earlier, they didn't picket the construction site of the new, and almost complete, Lions Gate Hospital, permitting other trades to continue working the site.

The steel on the south shore was poised on its falsework, a few days short of landing on Pier 16 where it would be safe. The ironworkers had struck the bridge at a time of maximum advantage. Colonel Swan, who was surprised that the men had left before landing the span, cautioned that "The bridge is reasonably safe, but there is a certain hazard involved. It is not a good position in which to

Drivers looked up nervously as the south-side steel sat gingerly on its falsework while the men were on strike. OTTO LANDAUER OF LEONARD FRANK PHOTOS, JEWISH MUSEUM ARCHIVES LF-36342

leave it."[230] The steel hovered over the road and the rail tracks leading to the old Second Narrows Bridge, causing people driving beneath it to look up nervously. They were probably asking themselves, *Wasn't it the falsework that collapsed on the north side a year ago?*

Although the CPR had issued a "slow order" to its engineers, the legs of the bent closest to the road had been boxed with steel, and protective netting had been hung below the steel to prevent any errant bits of construction material from dropping onto trains or cars, none of this alleviated the public's jitters. Their confidence might have been further eroded had they caught a glimpse of Don Jamieson sitting at the end of the partially completed span, which he did almost every day of the strike, worrying that it would collapse before it was landed. His wife Jean remembers that the strike made him sick, so that he was hospitalized at Burnaby General Hospital where he was, coincidentally, the chairman of the board. He was taking his responsibilities seriously.

Company officials commented to the press that they hoped for a last-minute settlement, though they would not comment on plans to exercise their legal options through an injunction. Other company executives, meanwhile, were working behind the scenes to try to get the men back to work. John Prescott had, a few days earlier, sent a telex to J.R. Downes, general secretary of the International Association of Bridge Structural and Ornamental Ironworkers based in St. Louis, Missouri. It read:

> THIS COMPANY HAS TODAY RECEIVED FORTY-EIGHT HOURS NO-
> TICE OF STRIKE FROM LOCAL 97 WHICH FOLLOWS A COMPLETE
> BREAKDOWN IN NEGOTIATION STOP WE ATTEMPTED TO PHONE
> YOU THIS MORNING STOP YOUR REPRESENTATIVE JOHN MARCUS
> HAS GIVEN US CONTINUING ASSURANCE THAT YOUR HEADQUAR-
> TERS WOULD NOT ALLOW WORK STOPPAGE WHILE ANY ERECTION
> PROJECT IN A HAZARDOUS POSITION STOP OUR SECOND NAR-
> ROWS BRIDGE PROJECT IS BEING CANTILEVERED TWO PANELS
> FROM FALSEWORK BENT TO SOUTH MAIN PIER STOP IT IS UNDE-
> SIREABLE THAT ERECTION SHOULD STOP UNTIL SOUTH MAIN PIER
> REACHED WHICH WE BELIEVE WILL REQUIRE TEN WORKING DAYS
> THIS BRIDGE CROSSES OVER A MAIN LINE RAILROAD AND MAIN
> HIGHWAY AND MOST UNDESIREABLE TO LEAVE THIS SPAN ON
> FALSEWORK FOR EXTENDED PERIOD OF TIME WHILE LOCAL 97 ON
> STRIKE STOP WILL TELEPHONE YOU MONDAY MORNING REQUEST-
> ING A POLICY FROM YOU IN REGARD TO THIS BRIDGE.[231]

That Downes failed to respond, and was not available by telephone despite repeated attempts to reach him, prompted Prescott to call Local 97 and ask for Norm Eddison. Before he accepted the call, Eddison asked Fernand "Fernie" Whitmore, president of the Local, to pick up an extension and listen in. They both heard Prescott ask, "Will you stay with us to complete that span?"

"I do not have the power to call off strike action," Eddison explained. "I'm just a servant, a hired man, but I will place your request before the executive committee of the union who will meet at their regular meeting in one hour's time."

"I do not want you to place any request before the executive board," Prescott responded tersely. "I have asked you, an officer of the union, and you have refused to co-operate. This is just fine," he concluded.[232]

The company now had all the ammunition it needed to instruct its solicitor, Lyle Jestley, to approach the court to seek an ex parte (only one side is heard) back-to-work injunction which was duly granted by Justice A.M. Manson. The wording of his order was clear:

> . . . an injunction is hereby granted enjoining the said Local 97, its members, agents, servants and every of them from striking and/or calling members of the said Local 97 on strike and/or withholding members of the said Local from their employment insofar as such strike affects the southern portion of the structure known as the Second Narrows Bridge presently under construction across Burrard Inlet until such time as the said southern portion of the structure is rendered entirely safe in the opinion of the Consulting Engineers for the British Columbia Toll Highways and Bridges Authority but not for a period longer than four clear days.[233]

Four days was the maximum an ex parte injunction was permitted under the new Trade Union Act, and it was estimated by a very few that it would take less than that time to build the steel out to Pier 16. But if more time was required, and that was likely given the engineers' ballpark of ten days, an application would be made to extend it.

The injunction would only apply to the south side of the bridge, where ironworkers were instructed to lay down their pickets and return to work. The picket signs on the south side were, in fact, picked up by Fernie Whitmore and conveyed back to the Local's office, after which Eddison phoned as many ironworkers involved on that side of the bridge as possible to inform them of the court's order. But then he started to read Swan and Prescott's affidavits which had been forwarded with the order, noting that they reported the extreme danger of the falsework legs being so close to road and rail traffic, and that an accident or an earth tremor might cause them to be broken or damaged. Small earthquakes were not unknown in the region. The danger was reinforced by Wooster, Swan's partner, in his affidavit: "That in view of the possibility that the bridge may be supported on a temporary structure for an indefinite period of time it could be considered hazardous, as all structures are during erection." Yet he tried to minimize the risk to the ironworkers by stating that "but to state that it is dangerous for workers is an exaggeration completely unwarranted and unjustified by the facts."[234] As well, John Prescott informed the International's representative John Marcus that the bridge was "safe enough," and that "The only thing that would affect it is an earthquake."

The conflicting opinions of whether the bridge was safe or not didn't sit well with Norm Eddison, who issued an immediate press release:

> Until such time as we are personally satisfied that William George Swan, consulting engineer, has exaggerated the hazard, and is in error, and an independent engineering authority has evaluated the suggested precariousness of the structure and that further, all steps have been taken to positively prevent and make impossible damage to falsework from vehicular traffic or trainwreck or, alternatively, no men to be allowed upon this structure when motor vehicle or train traffic passes under this span, an ironworker would tempt fate to resume work.[235]

Eddison, as well as the other ironworkers, obviously knew the risks; they were no more than they faced daily on this and many other structures. But it was a convenient bargaining chip that they were prepared to play out to the end.

The government, meanwhile, was biding its time. It was anxious to test the strength of its new labour legislation, but it didn't count on the determination of the ironworkers, whose inventiveness neatly sidestepped even the court's powerful order. Since the men were restricted from picketing, the only show of force left to them was to refuse to work, which they did en masse for supposed safety reasons, causing DB lawyers to consider applying for a contempt of court action and a suit for damages. Norm Eddison said in response, "I would personally advise our men that they might imperil their lives by working on the bridge,"[236] and that "I would rather rot in jail than order the men back on the job."[237] WCB even added its voice to the mix, writing to the union that in its opinion the bridge was safe, but Eddison pointed out that the WCB inspectors weren't professional engineers, a point that they had made themselves during the royal commission.

In an effort to manage the situation, and perhaps add some influence at the bargaining table, the Building Trades Council brought in Ed Kennedy, regional director of the AFL-CIO Building Trades Department for Western Canada. But far from adding a voice of reason, Kennedy merely inflamed the passions of DB when he accused it of having the "worst relations with its workers than any firm in the business."[238] This may have had a kernel of truth, given that the company was only now settling with collapse survivors for the loss of their boots and equipment a full year earlier. As well, the fact that the giant steel erector had cut the men's wages at the time of the collapse rather than let the shift run out in their favour, didn't ease the tension any. "We were swimming instead of working and that's why they cut the time," Bill Stroud said publicly about the contemptible action. The fact that DB never offered any satisfactory explanation suggested that he might have had a point.

Jestley, meanwhile, was applying to the court for an extension of the injunction, which was due to expire on the Saturday, two days hence. Ike Shulman of the firm Shulman, Tupper, Gray, Worrall & Berger, solicitors for the union, was simultaneously applying to have the same injunction set aside, a motion that would be heard by Justice A.M. Manson on the same day that both parties would appear before him to present their evidence of the bridge's safety. The judge offered that if enough evidence was put before him that the union had defied the court, he would have no option other than to rule on the matter—but the court would not initiate such a proceeding. Although Jestley wasn't ready to apply for a contempt ruling, he was ready to sue the union. He commenced the action on June 25, for unstated damages.

The hearing proceeded the next day. Representing the company was Lyle Jestley and in the union's chair was a young Vancouver lawyer, Thomas Berger, who had only just been entrusted the case by his colleague Ike Shulman. Berger was twenty-six years old, only two years at the bar, but already an accomplished lawyer. A year earlier he had defended, and won, his first murder case. Berger's labour roots were already deep. While attending university he had spent five summers working at a North Vancouver sawmill on the green chain as a member of the IWA, Local 1-217, and his wife's stepfather, Cliff Worthington, was a business agent for the Carpenter's Union. It was Worthington who had suggested to the young law graduate that he article for the firm which had a number of trade union clients, one of them being Local 97. The case would be a defining mo-

Tom Berger, lawyer, defended Local 97 against a draconian piece of labour legislation. PHOTO FRED S. SCHIFFER

ment in Berger's career, not only for its importance in trade union circles, but also for the confidence and poise that he would exhibit under fire.

Berger was appearing before one of the most cantankerous justices that a lawyer could possibly face. Justice Manson, whose epithet "the hanging judge" was probably well earned for reasons other than the obvious—capital punishment was still in effect at the time—had once been the provincial minister of labour. That he wore his dislike for trade unions on his sleeve was intimidating enough, but when added to that the annoyance he openly exhibited for the young upstart, it was a challenge Berger would meet, never forget and later write about:

> Justice Manson was in his mid-seventies, and widely regarded as being anti-labour. He was a man who had been active in the political life of British Columbia before going to the bench, and he had served almost a quarter of a century on the Supreme Court. A gifted lawyer, he was given to haranguing those who appeared before him: accused persons, witnesses and lawyers. He was also a fervent Presbyterian. He urged the men and women he sent to jail to turn to religion. In divorce cases, he upbraided peccant spouses. He chastised witnesses whose evidence contradicted his own views.[239]

Berger's first act was to question the justification for the ex parte injunction, given that none of DB's legal rights had been infringed:

> I wasn't, of course, present when this injunction was granted, but it appears that the only justification whatever for it is the suggestion that there is some authority to enjoin a legal strike under the Trade Unions Act, and my submission is there is no such authority. It rather restricts the procedure and the use of ex parte injunctions but does not confer

any substantive right to enjoin a legal strike. It is restricted to actions, unlawful actions, relating to illegal strikes.[240]

Justice Manson described the Act as "novel legislation," and elaborated:

> It is new, passed for the very purposes that are disclosed in the section itself. It is legislation that some might regard as extraordinary which the legislature saw fit to pass . . . The purpose is obvious. It is to take care of actions on the part of a trade union which may endanger public order or which may prevent substantial or irreparable injury to property . . . An act which is otherwise legal may be an act that undermines public order or that may result in substantial or irreparable injury to property. It doesn't follow at all that the act must be an illegal act in the ordinary sense.[241]

Although Berger believed that the union had the law on its side, he also recognized that the authority of the Supreme Court was final, and "that if a superior court judge, even erroneously, makes an ex parte order, it must be obeyed."[242] After Justice Manson had Jestley read aloud the affidavits of both Swan and Prescott, Berger argued that "if this bridge is in a dangerous condition then it is just as dangerous if the men are on the bridge as if they are on strike, and I submit these affidavits, far from setting up anything like tenable grounds for an injunction, set up just the opposite."[243] It was an argument, however, that would fall on deaf ears.

Justice Manson sided with the bridge company, allowing that their request for the span to be completed up to Pier 16 was "common sense." Berger later wrote about the decision:

> What the judge was saying might, in the abstract, have been reasonable or prudent. But the issue before him was whether the ironworkers employed on the bridge had the right to strike; whether the courts could order back to work men who were lawfully on strike. Judges do not sit to dispense abstract justice according to what seems to be "the common sense of the thing." They can only enforce the law. The legislature can— and from time to time it does—pass legislation to put an end to a legal strike. But judges have no such power.
>
> Now Justice Manson told the union to order its members to return to work on the bridge until the southern approach had been

completed to Pier 16. Dominion Bridge had not asked for such an order. The judge was making up remedies as he went along. Clearly it rankled that his ex parte order had not resulted in the men going back to work. The judge made this new order as he left the bench. As far as our motion to set aside the ex parte injunction and Dominion Bridge's motion to extend the injunction were concerned, they had not been dealt with.[244]

Berger's firm immediately launched an appeal on grounds that the affidavits contained misrepresentations, that the strike was legal, that the evidence disclosed no threat to public order and that Justice Manson had erred in "holding that Section 6 of the Trade Union Act enabled him to enjoin a legal strike."[245] George North, editor of the staunch unionist paper *The Fisherman*, condemned the judge's action as well when he penned an editorial entitled, "Injunctions Won't Catch Fish nor Build Bridge." North wrote that the court's judgment was partial and prejudiced, an opinion that had Justice Manson mulling over whether he should charge the radical editor with contempt.

"It has been held that the courts are not above criticism in the press," Manson noted, "but to suggest partiality and prejudice is quite another matter."[246]

The government was now becoming impatient, prompting Premier Bennett to release an engineering assessment that he had commissioned at Norm Eddison's request to speak to the potential danger. Prepared by an independent Vancouver engineering firm, Phillips, Barratt and Partners, it read in part:

> The structure as presently supported by falsework bents is in safe condition for all normal risks of bridge construction.
>
> However, acts of God such as a major earthquake might cause severe damage to the structure.
>
> We would therefore recommend that the soundest way of minimizing such risks would be to continue construction to the main south pier which is the next permanent support.[247]

Local 97 business agent Tom McGrath commented that the study didn't support DB's claim that the falsework was unsafe, and that the "report has confirmed my opinion that many things could happen to that bridge."[248] He did, however, expect that the men would return to work pending Ike Shulman's request to have the injunction set aside. After Manson's second edict requesting the union to order its men back to work on the south side, Norm Eddison stated that the court order "will be obeyed in its exactness"[249]—though he made no comment when asked if that meant the men would be returning to work.

When the Monday dawned and the thirty-one ironworkers, including survivors Norm Atkinson, Jim English, Lou Lessard and Bill Stroud, failed to show up for work, nobody was surprised—least of all, perhaps, Norm Eddison and Tom McGrath. Although they had phoned the men to inform them of the injunction in compliance with the court order, the fact that the ironworkers were without a collective agreement meant that it was up to each of them whether they reported for work or not. Eddison pointed out that the matter was now between the men and the company, and disavowed any knowledge of what was happening on the bridge, advising the press that he didn't want to go near it for fear he might be accused of "hindering the men from returning."[250] He then offered that he only knew the whereabouts of one man whom he had called. Lou Lessard was busy having a forty-two-centimetre stainless steel rod removed from his leg, which had been bolted to his shattered bones after the collapse a year earlier.

The ironworkers strike had many unions in the province wondering where the BC Federation of Labour had been when the legislature had passed Bill 43. They hadn't challenged it, leaving some to call their inaction "spineless" and "weak." One delegate at a Nanaimo convention commented that the federation should have taken some action, even if it meant, "tying up the province for a stated length of time."[251] But the BC Fed were now making up for lost time, firing off missives to Prime Minister John Diefenbaker, federal Justice Minister E. Davie Fulton and provincial Attorney General Robert Bonner, asking them to remove Justice Manson from the case.

The company, meanwhile, still believing that "common sense" would prevail, served eighteen ironworkers with copies of the injunction, the WCB letter as well as Premier Bennett's wire to the union leaders stating that the bridge was safe. It had no effect. Several ironworkers commented to the press that they were talking to the wrong people, that they should be talking to the union, from which they themselves were awaiting direction.

While the union and its members were discussing the order, Lyle Jestley, who now considered the men to be in contempt, was back in court petitioning for a writ of sequestration, a vehicle used by the court to seize assets pending resolution of its judgments. Justice MacFarlane advised him that he could now inform the union that an application would be made before Justice Manson on Thursday, July 2, for a writ authorizing the sheriff to seize assets for failure to obey the court

Survivors. Back to front: Bill Stroud, Norm Atkinson, Lou Lessard, Charlie Geisser, Colin Glendinning and Bill Wright. PHOTO BILL CUNNINGHAM, *THE PROVINCE*

order. The ability to sue the union was another fringe benefit that the government had conveniently added to the Trade Union Act.

Talks between management and the ironworkers were now on hold. The union wanted individual meetings with each company, and specifically stated that they wouldn't meet with anyone representing DB except for an employer's committee. They also wanted all future meetings to be held at the union hall, otherwise negotiations would have to be conducted by mail. Despite the vitriol, R.K. Gervin, negotiator for the Structural Steel Association of BC, was still optimistically suggesting that talks had not broken off completely.

Meanwhile, at 4:31 p.m. on July 2, bonded investigator Charles Reid arrived at the union's offices at 111 Dunsmuir Street to serve Fernie Whitmore, Tom McGrath and Norm Eddison with the court's order. The three would appear before Manson a week hence, on Thursday, July 9, when Jestley would suggest to the court that the reason none of the eighteen men had shown up for work was that the union had not told them to. Each of the three men then lodged counter-affidavits swearing that they had complied with the court order by phoning all the relevant men and telling them of the decision, but they also admitted to advising them that, upon advice from counsel, the judge's order was "contrary to the law."[252]

After Jestley argued that the failure to comply with the ex parte injunction was a criminal and not civil contempt, as it involved "public injury of offence," he advised the court that he wasn't sure that the men understood the situation they were in and that "I think particularly these men should have been told, if they weren't, that there is a possibility under all the authorities that they could be cited or brought in for contempt because I do not think they have had that chance."[253] He then suggested that the union may have pressured its members.

Although Justice Manson offered to have the men brought to court so that the impact of their actions could be fully explained to them, Jestley expected that that wouldn't be necessary. Manson then launched into a long tirade, stating that his order would stand until it was overturned by an appellate court: "The situation has all the appearance of an absolute defiance of an order of the Court. We have in Canada what is known as the rule of law. We live by that. The rule of law must be maintained. If you have anything to say, Mr. Berger, now is your chance."[254]

Berger immediately objected to Jestley's inference that the union had brought "improper pressure to bear" on the men, given that he had admitted to having no such knowledge of that happening, but Manson just as quickly interjected, stating that the onus was on Berger and not on Jestley to prove that there had been no defiance of the court, despite the affidavits of the three leaders swearing otherwise. Berger reflected on the situation in his 2002 autobiography, *One Man's Justice*:

> What Justice Manson was up against, was this: He had made an order requiring the union's officers to direct the men to return to work, but the ironworkers themselves were under no legal obligation to return to work. As for the union's officers, they had to order the men back to work, but once they had done so, they had no further responsibility under the court order if the men did not return to work.[255]

As the case progressed, Berger had been spending more and more time in the law library, reviewing old cases on sequestration, when he stumbled upon one of the most fundamental documents influencing both common and constitutional law in the Commonwealth, the Magna Carta. To Justice Manson's surprise, and likely to his annoyance, Berger launched into his defence referencing that venerable document:

> My submission is that for your Lordship to order a writ of sequestration against the union in this particular case, if your Lordship should find that there has been disobedience to your Lordship's order, and I submit that no such finding can be made on the material, but if your Lordship should so find, your Lordship cannot issue a writ of sequestration because to do so would be contrary to our whole constitutional law, and I refer to the Magna Carta in support of that submission.[256]

Although the Magna Carta had been written in 1215 to define the rights of the king following arguments between King John and the Barons of Runnymede, it was still considered relevant. Ratified as English law in 1297, chapter 15 of the document decreed that "No free man shall be distrained to make bridges."[257] Berger expounded on the context of the statement:

> The proviso there refers to the obligation resting upon certain feudal tenures to make bridges as part of the services incidental to the tenure, part of the services owed to the Norman kings in respect to the grant of land comprised in the tenures, and what had happened was that King John had attempted to extend the obligation to make bridges to free men and to communities that had not been obliged to do so under the

terms of their grants from the Norman Kings, and, it was by reason of that mischief that the Barons extracted from King John in 1215 this decree that free men should not be distrained to make bridges except those in the tenures in the lands that had at old times been obliged to do so.[258]

Manson stated that there was much later law than Magna Carta, advising Berger that he had heard that argument fifty years earlier in university. "Start at the other end," he implored, "and if you get back to Magna Carta, that will be all right, but start with something current."

Berger replied, "I am coming to those cases, but Magna Carta is current. It is as much in force in British Columbia today as it was in England in 1215. It is part of the constitution of every one of her Majesty's realms and British Columbia is one of them." Berger, now at his wit's end, recognized that the judge was not going to be swayed by his arguments, and that he had better start speaking to the record rather than the court in order to lay the basis for his intended appeal.

Although Manson accused the young lawyer of playing to the gallery, Berger insisted that he wasn't, affirming that "No statute of this province can infringe the rights or liberties granted free men by Magna Carta."[259]

The judge begged to disagree, advising Berger that Magna Carta could be repealed by both federal and provincial legislation, but the introduction of such an unusual defence made him pause. Rather than face a lengthy constitutional debate, he elected not to sequester the union's assets. He did, however, order that the eighteen recalcitrant men appear before him on Friday, July 10, for examination, although he also instructed Sheriff Eddie Wells to be at the bridge site at 8 a.m. on that day and if any of the men showed up for work to excuse them from the court.

Berger strenuously objected, arguing that the officers of the union, the only party that had been enjoined by the injunction, had complied with the court, and just because the men had not obeyed the union did not mean that they were in defiance of the court. Berger wrote about his frustration:

> I thought I had the answer to the injunction; then I thought I had the answer to the judgment requiring the union to order the men back to work; then I thought I had the answer to the company's application for sequestration of the union's assets. But each time the judge veered off in a new direction. The judge was inventing his own procedure, because none in the books suited his purpose.[260]

The sheriff dutifully set off to serve the men, but was only able to find eight of the eighteen. One of those, Lou Lessard, was still recovering from his operation to remove the metal rod from his leg. Al Snider, one of the men who was subpoenaed, remembered that "there were some subpoenas issued for me, Eric Guttman, Johnny Phillips . . . there was a couple more . . . but anyway we got a phone call from Norm Eddison to leave the house." Despite Manson's instructions for union officials, plus Berger, not to talk to the men or face contempt charges, Norm Eddison impudently defied the judge by warning the men that they were about to be hauled into court. He was not about to see any of his members go to jail if he could help it.

On the Friday morning, when the list of the eighteen men was read in court, none of them were present, causing the now livid judge to issue bench warrants for their arrest. Despite Berger's objections and three attempts to "make a number of things clear for the record,"[261] the justice instructed the sheriff to have the men in court on the Monday to face contempt charges. Admonishing the court, he fumed:

> I have been rather impressed with the fact that there has been an attitude of defiance, not by the workmen. I know workmen, I came from them. I was one of them. I wasn't born with a silver spoon. I should not introduce a personal attitude, but I was Minister of Labour for seven years, and I was their solicitor for many, many years in a very substantial area. Workmen are for the most part men of common sense, reasonable individuals. Unfortunately, sometimes they are dragged into situations which are embarrassing to them. Now, in this particular case it is my opinion the workmen should consult independent counsel.[262]

Not only was the judge recommending independent counsel, but he then specified that the counsel should be "senior," "ripe" and "experienced," an obvious slight to Berger who was prohibited from even talking to the men under threat of contempt.

On Monday, July 13, only seven men appeared in court, the others having conveniently slipped the sheriff's noose. Justice Manson instructed the sheriff to continue to search for the missing ironworkers and have them in court by 2:30 p.m.:

> . . . [it looks like] a concerted evasion of service . . . I certainly shall not take a lenient view if they insist on evading service. We will be building

bridges for some time to come and I do not intend to have a conspiracy to evade the orders of this court . . . As I see the situation at the present time there was no bona fide compliance on the part of the union to obey my direction to call the men back to work.[263]

Berger piped up: "May I say that the evidence before your Lordship is all to the contrary."

Manson snapped, impatiently: "Mr. Berger, please do not tell me that. I have read the evidence and I do not want to be drawn into an argument with you about that."[264] Berger later wrote in defence of his position:

> But that's what you do in court. You have an argument. Then the judge decides. We had had a series of discussions, but no argument . . . The judge said he had read all the evidence. What evidence was there that the judge could have read? All the court knew, on the record, was that the men had been ordered to go back to work and had not done so.[265]

The court was adjourned until 2:30 p.m. at which time Manson expected to not only see the eighteen men who had been served, but all thirty-one of the south-side ironworkers in court. After examining them, he intended to make his ruling.

When the proceedings opened later that afternoon, another face was present at the defendant's table. The ironworkers had indeed hired "senior" counsel to represent them. Tom Hurley, their representative, then in his mid-seventies, was according to Berger, "as courtly as Justice Manson was rude. They were old antagonists. The judge was not pleased when Tom Hurley appeared for the men."[266] Hurley also happened to be one of Berger's mentors, and had the unfailing wisdom and battle scars of experience. Berger remembers that he was a great sounding board for young lawyers and that he never got angry at anyone, even judges who had ruled against him.

Hurley, with the union's help, had managed to round up all eighteen men who had been served—the other thirteen were still missing—and despite him questioning the judge as to why he and the ironworkers were even there, Justice Manson assumed a prosecutorial role as he prepared to examine them. Berger's objections were also overruled, as was his request for clarification whether the proceeding was a criminal or civil one. The difference was significant, given that the men could incriminate themselves were it criminal.

"I will determine that later," the judge snarled, as he charged into the examination of Eric Guttman, the first of his victims.

Manson was determined to expose the union's business agents, and endeavoured to trick Guttman into disclosing that they had called him after the court had issued its order restricting them from doing so. Guttman stuck to his story that he had only received one phone call and that they had only told him to report to work on the Monday. The judge persisted, telling Guttman that his testimony was in direct contradiction to that provided by Eddison and McGrath, causing Berger to bounce to his feet to voice his objection: "There is no evidence indicating your Lordship's statement to be correct."

"Just sit down," the justice commanded. "I am doing this. You keep your seat." [267]

Manson pressed on, causing Berger to ask him to read aloud the conversation contained in the business agent's affidavits, which stated that they had also told the ironworkers that, according to counsel, the proceeding was "contrary to the law." The justice's immediate response was to threaten Berger with contempt for interrupting him. The gallery, now packed with about seventy-five ironworkers, broke out into raucous jeers. They knew that the proceeding was a sham. The sheriff shouted, "Order in court!" causing the judge to turn to the gallery and plead for order as well. Ernie Duggan, who had lost his son Kevin on the bridge and who would lose another son, Gordon, on another project a couple of years later, shouted over the rabble: "Rah, rah, rah for British justice" and "You said it, bud! Keep talking, lawyer." The sheriff quickly showed him the door.

The judge then asked Guttman directly if he had been told by the business agents that the proceeding was "contrary to law." He answered, "Yes," claiming that the reason he had denied this part of the conversation earlier was that "There was so many things I remember that was told." Further questioning about why he had obeyed the order to strike but not to return to work, elicited the following response: "it is a free country and nobody can force me to go to work to build a bridge if I don't want to." After several more attempts to inquire why he hadn't returned to work, the judge abandoned his examination and invited apprentice John Phillips to the stand. He then launched into another interrogation, causing Hurley to ask for the second time why he and the men were there, as he had heard no case against them. After Manson responded evasively, Hurley replied cheekily that "I thought that blarney was a monopoly of mine."[268] He was probably one of only a select few who could get away with such insubordination before the cantankerous elder.

Following Phillips came Al Snider and Isaac Hall. When Manson pointed out to Hall that a Vancouver engineering firm had reported that the bridge was safe, Hall responded, "But can we always believe these engineers? They told us a year ago that the bridge was quite safe. Eighteen of my pals got killed on that bridge a year ago."[269] The message imparted by three others including Bill Wright was much the same. Manson gave no courtesy to Wright for being a survivor, treating him with the same disregard as the rest of the men. In the end, none of them would admit to the business agents telling them anything more than to report for work and none of them would report for safety reasons. The men were excused after their examinations and the court was recessed for the weekend.

While the next two days should have offered Manson an occasion for sober reflection, he used the time to strategize instead. On the Monday morning he informed the court that he had decided to invite the attorney general, through counsel, to participate, offering that it was "unfair that the plaintiff should carry the burden."[270] At 2:30 p.m., M.M. McFarlane was introduced as counsel for the attorney general, but as he needed time to come up to speed, the court was adjourned until Thursday, July 23. Then, McFarlane's first order of business was to introduce an order calling on Whitmore, Eddison and McGrath to appear before the court to "show cause why they should not be committed for contempt of this court in disobedience of the Court's Order of the twenty-sixth of June in this action." Berger objected, stating that there was no evidence that "would justify the Order," and that as the only motion before the judge, the writ of sequestration, was a civil one, it was improper to engage the attorney general. His objection was summarily dismissed.[271]

When the proceedings resumed on July 28, Justice Manson had a surprising announcement. On the preceding Saturday, July 25, he had received what he called "fan mail," though the correspondence was far from complimentary. It was addressed, "Dear Dictator," and threatened that "Unless you immediately issue a court injunction ordering all Steel Companies to give us our pay increase, and other demands, you will be filled so full of holes, you will look like a Swiss cheese." A postscript also warned him that "in case you jail any of the men involved, prepare yourself for civil war."[272] It was signed, anonymously, "Steelworker," leaving Berger with the impression that it was not an ironworker who had penned it, but a steelworker who belonged to a different union. But as Berger would later write, the fact that he had had to address the issue at all, which was wholly unrelated to the contempt charge, indicated how far off the rails the proceedings had fallen. At the end of the day Manson upheld the order despite Berger's objection that no contempt had been shown.

On July 30, the judge presented his Reasons for Judgment, finding the union and its officers in contempt. The union's two business agents and its president, Norm Eddison, Tom McGrath and Fernie Whitmore respectively, were arrested and carted off to the Oakalla Prison Farm. Earlier, Eric Guttman had been arrested, but released after paying a $100 fine. That ironworkers had had a long tradition of being on the business side of cold steel bars in times of trade union stress was probably not lost on these men. In fact, it was to be something of a badge of honour, evidenced by the standing ovation they later received at a BC Fed rally.

Eddison, McGrath and Whitmore were fined $3,000 each, an impossible sum for most, and in default, imprisonment for one year. Eddison managed to pay his own fine while the BC Fed coughed up enough to spring McGrath and Whitmore. Pat O'Neal, secretary of the BC Fed, expected that the International Union would cover the union's $10,000 fine. George North, editor of *The Fisherman*, whom Justice Manson had decided to charge with contempt after all, was also fined $3,000 and sentenced to thirty days in jail, though he said he would fight the charge on appeal.

The ironworkers' strike was only one of many happening that summer, which saw nearly 40,000 men and women protesting everything from poor wages to poor benefits to a thirty-five-hour week. Although the injunction obtained by DB was the first of about a dozen offered to various companies by the courts, courtesy of Bill 43, the ironworkers' strike seemed to attract the most press. Perhaps it was because everyone could see the inactivity on the bridge, a vital piece of infrastructure that was long overdue, or maybe it was just due to the controversy generated by the court. Whatever the reason, the strike was now into its seventh week, but by the end of the first week in August the two parties were separated by only twenty-four blocks, representing the physical boundary that started the wage clock ticking for ironworkers travelling in Greater Vancouver.

The company wanted the eastern boundary to be at Sperling Avenue while the men wanted it twenty-four blocks west at Boundary Road, the difference representing about half an hour's pay. Despite being so close, the men rejected the company's latest offer by a subsequent vote, but after a few days were presented with a "secret proposal," granting them the Boundary Road divisional line. They accepted, pulled their pickets after fifty trying days and promised to resume work as soon as possible. The union had achieved almost all of what it had set out to obtain. Not all unions, however, viewed the ironworkers' fifty-seven-cents-per-hour wage increase with envy. The Operating Engineers working for DB, though offered the same increase, rejected it as too low. They were about to strike for an unheard of sixty-four-cents-per-hour increase. A few construction companies were also unhappy with the increase, an additional cost that convinced some of them to take their business out of province, but with almost $800 million in construction work in BC at the time, the ironworkers were unfazed by the handful who decided to jump ship.

The case against the ironworkers eventually ended up in appeal court, Berger representing the union and McFarlane upholding Justice Manson's decision. It was heard by Chief Justice A.C. DesBrisay, Justice H.W. Davey and Justice J.A. Coady, who unanimously found that there was no justification for Justice Manson's ruling of contempt. They reversed the order and instructed the court to refund the $19,000 in fines. Not only did they set aside Manson's ruling, but Berger was also advised by Justice Davey that he shouldn't have been rebuked for rising to object during Guttman's examination, nor threatened with contempt.

"Justice Manson was no hypocrite," Berger later wrote. "He hated unions. He used his position on the bench to impede and frustrate them when he could. But he hardly took the trouble to pretend that he was dispensing even-handed justice. There were no long-winded rationales for his judgments. He was out to get you, and he did."[273]

The consequence of Manson's behaviour, which seemed so shamelessly inappropriate, was that the government passed legislation making it mandatory for superior court justices to retire at the age of seventy-five. Manson was already past that age when he took on the ironworkers case and he would be a lofty seventy-eight before he was forced to step down. The new act was eponymously nicknamed "The Manson Act" by those familiar with the querulous judge.

While the three justices had been hearing Berger's appeal, in another courtroom, the same judge's contempt charge against George North was being upheld. North's lawyer John Stanton couldn't remember any other Canadian editor being jailed for contempt except for John Robson, a young New Westminster newspaperman who in 1862 was thrown in jail by "hanging judge" Matthew Baillie Begbie after he made an unsubstantiated accusation that the judge had accepted a bribe from land speculators. Robson later become British Columbia's ninth premier.

12

Closure

*Every man on the job has a tremendous thrill of pride and
accomplishment when a bridge is completed. Each one feels that
he is an integral part of it—as indeed he is.*

–DON JAMIESON, P. ENG.[274]

The Exclusive Order of Prudent Penguins was a unique club. There were no rules, there were no meetings, but all of its members had one thing in common: they had all taken an unplanned dip in the inlet but had been saved from drowning by wearing a buoyancy device. Norm Atkinson was a member. So was Jim English, Bill Stroud, Lou Lessard and a dozen or so others who had fallen into the water with the collapse. Bob Dolphin also became a reluctant affiliate of this elite fraternity during the erection of the south-side steel, and he would be that chapter's last member.

He recalled the day that he joined the club, the day that he climbed to the top of Pier 16: "We had a steel ladder on there, and when the tide ran it would swing aside, so you'd grab onto the bottom and your weight would bring it vertical, and I was climbing up with a transit in my hand and the tripod on my shoulder, and I grabbed the top rung and it came off and I fell." Dolphin surfaced, still clutching the transit and tripod, the seven-metre drop dampening his clothing and instruments but not his spirit. Fortunately his small tender, which was tethered to the bottom rung of the ladder, was only a few swim strokes away and he quickly struggled aboard.

By the middle of September the men were now well into the swing of steel erection on the south side. Pier 16 had been reached and the steel was now safe. Don Jamieson breathed a well-earned sigh of relief. His next challenge would be

Bob Dolphin, erection engineer, with level, the same job that John McKibbin performed before the collapse. COURTESY BOB DOLPHIN

to erect the long clear centre span consisting of the two 115.5-metre (north and south) cantilevered arms plus the 104-metre suspended middle span known as the "drop in." No falseworks would be required for this stage of construction, which would rely solely upon cantilevering to push the steel a lengthy 167.5 metres from each side of the inlet.

Cantilevering such long spans successfully required careful planning. The spans would be under increasing tension as they moved out over the water, daily gaining weight, with the anchor assemblies of Pier 17 on the south side and Pier 14 on the north side being the only restraints that would prevent them from collapsing into the inlet. It was only when the south-side span met the northern span in the middle of the inlet, in a process called "closure," that the stresses on the steel would change. Closure day would be a relief to all. But before the north-side steel could even be started, the southern span had to be complete. Traveller No. 1 could not be moved to the north side before that occurred.

The first two bottom chords, two of the largest pieces of steel ever employed in a western Canadian construction job, were delivered to the site on Thursday, September 17 by rail barge. It would have been too risky to move each of the approximate 96.5-tonne, 24-metre members by truck. Before these two immense pieces of steel could be launched from the "jumping off" Pier 16, two diagonals had to be dropped from the fixed steel above to hold them in place. Once the

"Would the two cantilevered arms meet as anticipated?" This was what Don Jamieson was anxiously contemplating just days before closure. PHOTO JIM PRATT

bottom chords were erected, the lighter vertical members (plumb posts) were installed and over that the top chords were lowered onto the whole panel. The three bracing systems followed to make the structure rigid, and two to three panels behind the front end, floor beams and stringers were installed. Although the forms were also installed for the concrete roadbed, the pour would not occur until after closure when the completed span could support the additional weight.

Donny Geisser, who had assisted with the dive recovery during the collapse, was now on the bridge to replace the late Frank Hicklenton, who had been his dad's signalman. Donny's father Charlie, although still sore from the chest injuries he had sustained during the collapse, was back in the cab of the traveller, engaging and disengaging the gears of the huge spools that picked up and dropped the massive pieces of steel to the raising gang below. Donny was working with a familiar group of men, the same gang that had been erecting steel at the front end on the north side before the collapse, only missing Joe Chrusch. Art Street was in his place.

Donny's buddy David Milne (a.k.a. "Davy Crockett"), was also on the bridge, but he was in the bolting-up crew. Although Milne had been on the Peace River Bridge during the Second Narrows Bridge collapse, he was soon to have his own moment of apprehension at the Narrows:

> I was inside a top chord on the front end, and it was my turn to go inside, and all of a sudden—*bang!*—the whole bridge shook and dead silence. I hollered at my partner, but he's gone, wasn't there . . . and I waited inside for a few minutes . . . It hasn't fallen down so I got out on top to see what had happened and the swing line on the bull wheel that swings the boom had broken. The boom swung around and crashed into the moonbeam.

It was a relief, one that perhaps would not have registered as strongly had the bridge not had a history.

Steel erection had only been underway for a couple of months when it looked like another shutdown was inevitable. Members of the Hoisting and Operating Engineers Union Local 115, all working for DB, had voted twelve to six in favour of a strike after being offered the same per-hour wage increase negotiated by the ironworkers during their strike. They rejected it in favour of their original demand, a sixty-four-cents-per-hour pay increase that would equalize their wages with those earned by other operating engineers in other branches of the construction industry. Although DB would have only two weeks to agree before

As the steel was nearing closure, the men erected a record 192 tonnes of steel in just eight hours. "Nobody was driving them," said Don Jamieson, "they did it because they wanted to." PHOTO JIM PRATT

the men would walk, the men's demands were settled quickly; the company could ill afford another delay, especially with the Toll Authority breathing heavily down its neck.

In other union activity, representatives from the International had set up in Vancouver on the fifteenth floor of the Georgian Towers hotel. They had gathered enough information to charge, under their constitution, Fernie Whitmore, Norm Eddison and Tom McGrath for organizing an unauthorized strike. It had been said that the International wanted to rid itself of militant leaders, but the three accused had also not ingratiated themselves by vehemently disagreeing with the "sweetheart deal" the union executive had struck "for work at Kitimat, and for the 1956 province-wide agreement."[275] Although the three's defence revolved around two mysterious telegrams authorizing the strike, their argument fell flat when they could not produce the documents. The messages had been phoned in and not delivered, and the International denied sending them.

Gay Borrelli, third vice-president of the International, conducted the trial where all three were found guilty. The men would not only pay with their executive positions, but also with their union memberships, and when the decision was finally rendered on May 19, 1960, the assembly erupted in a near riot. The controversial decision divided trade unions across the province. Pat O'Neal, secretary of the 120,000-strong BC Fed, stated that the three had organized "one of the most successful strikes conducted in this province for a good many years"[276]—but as they had done so without the blessing of headquarters, the International considered it a serious breach of protocol.

Just before the International announced its decision, Tom McGrath angrily tendered his resignation, stating that the trial and the International's refusal to hear all of his defence, was a "vivid example of international gangsterism."[277] Despite losing both his executive position and his union ticket, McGrath would not be silenced, but the next time he would surface would be a much greater threat to the International than leading his members into a victorious strike.

The trial also presented evidence, by way of another telegram sent by John Prescott to the International during the strike, that the government may have interfered by ordering DB to seek an injunction to force the men back to work. The telegram read, in part: LOCAL 97 ARBITRARILY AND IRRESPONSIBLY CEASED WORK WHILE BRIDGE (SOUTH END OF SECOND NARROWS) IN HAZARDOUS POSITION, AND AS A RESULT THIS COMPANY WAS FORCED BY CUSTOMER TO OBTAIN INJUNCTION YESTERDAY.[278] Although Gaglardi fervently denied any involvement, suggesting that the word "customer" didn't necessarily mean the government, Eddison was adamant that there had been collusion between the company and the Toll Authority.

Opposition leader Robert Strachan thought so, too, as did two of his CCF colleagues: member for New Westminster Rae Eddie, and member for Burnaby Gordon Dowding. Dowding went further when he suggested that the attorney general should be investigating the act as an encroachment on civil liberties. Although Labour Minister Lyle Wicks defended the government by denying the accusation, he also stated that DB's action was "improper and highly irresponsible."[279] Wick's sympathies, however, did not extend to covering the union's $12,351 legal bill for defending itself against the giant steel erector in court, despite the union winning in appeal court. Ditto that for Attorney General Bonner, who suggested that the "proceedings were between private parties and the disposition of costs is a subject of court order. It is not an area in which we can intervene."[280]

Not many months after the trial, Norm Eddison would suffer a nervous breakdown. The strike and the loss of his membership and position in the Local was only the catalyst; his real pain had been the visits he had had to make with the police to the widows of his friends on the night of the collapse to inform them of their loss, and then being a pallbearer at nine funerals. He would soon recover, however, and return to the woods where he had begun his working life.

Despite delays caused by the ironworkers strike, steel erection was now moving quickly, and by the third week of November the south-side steel was only thirty-seven metres short of completion. That the men had made record time was largely due to cooperative weather. Where typically nine days were lost each October due to rain, only two were missed in 1959. It would only take another seven days to complete the south-side span before the men and equipment would move to the north side. There would still be a delay, however, due to the estimated one month it would take to dismantle, move and reassemble the traveller on the north side. On November 26, Bill Cunningham of the *Vancouver Province* photographed the last piece of south-side steel being lowered to the raising gang. Norm Atkinson, once again in his element, can be seen dangling fearlessly from the member, one leg kicking the air almost sixty-three metres above the inlet, the other with a toe-hold grip on the fixed steel.

While Atkinson and his colleagues were busy moving their equipment to the north side, the provincial Highways department was building the connecting section of the Trans-Canada Highway between the Second Narrows and Port Mann bridges (Port Mann under construction). The Upper Levels Highway connecting the bridge to West Vancouver was also underway, though experiencing its share of construction delays and federal compensation disagreements. The most difficult stretch between the Second Narrows and Port Mann was unarguably Burnaby's famous sawdust highway, stretching for almost a kilometre through the boggy reaches at the south end of Burnaby Lake. The sawdust, laid out over the proposed roadbed, was enough to heat the average home for as long as three hundred years. It was expected to absorb sufficient moisture to permit trucks to drive over the swampy surface, dumping as much as a metre and a half of sand over the course. Sand was also injected twelve metres into the bog at five-metre intervals through metal pipes which were driven, filled with sand and then removed. The weight of the sand over the sawdust was expected to squeeze the water up the sand columns, through the sawdust cap and into roadside ditches where it would drain away. With the water removed from the subsurface, the roadbed was anticipated to drop between two and three metres in elevation after an eight-month period of compression. The road would then be built up and surfaced.

The 19.5-kilometre, eight-lane route between the two bridges would end up costing about $20 million, almost as much as the bridge. Over two hundred homes had to be expropriated, but perhaps nothing was quite so problematic as the displacement of the resident muskrat population near the shores of Burnaby Lake. To the department's annoyance, the hardy little mammals moved into the cozy sawdust layer until they were evicted from there as well.

Just north of Burnaby Lake, the maze of southern approaches to the bridge was also in the final stages of completion, the accessible portions of which were to be open to the public on November 30. A month after that, the south-side steel was complete except for cleanup, and the north-side steel was nearing the same position it had been before the collapse. This time there would be no mistakes with the falsework. Extra pains were taken to secure it. Not only were the calculations worked, reworked and checked, but also a large tremie apron was poured around the base of the N4 piles to ensure that the pile nest was bound together as a single cohesive unit. Bob Dolphin even donned diving gear to inspect the forms before the tremie was poured. And although floor stringers were again used for the upper grillage, this time they were milled level before steel stiffeners and steel diaphragms were installed.

Don Jamieson advised that the steel erection was right on schedule and that he expected the bridge would be open for business no later than September 1. "Except for the actual joining of steel, the worst is certainly over," he said, after advising that only 150 metres separated the two sides.[281] But no sooner had he made this declaration than the north-side progress hit a snag. It was discovered that a portion of the cement used to construct the N5 pile nest required to support the next set of falsework had eroded since it was laid two years earlier. Work on the bridge stopped immediately and twenty-five men were laid off pending tests, but a week later DB recalled the men and resumed work, stating that the tests had concluded that there were no problems with the pile nest. By the beginning of March the bridge was poised over the N5 falsework, the same position it had been in on June 17, 1958, the day of the collapse. It was now only days away from landing on the "jumping off" Pier 15.

The imminent completion and inevitable naming of the bridge was distressing the new president of the Burrard Inlet Tunnel and Bridge Company, Murdo Frazer, who was also Reeve of the District of North Vancouver: "I intend to tell Mr. Gaglardi that we will be keeping the name Second Narrows Bridge . . . If necessary, I will insist. We've had the name since 1923 and I wouldn't like to see Mr. Gaglardi steal it for his new bridge."[282] Gaglardi, however, recognizing that the old bridge company had a copyright on the name, advised that he would be announcing the name of the new bridge soon.

Perhaps it was not the name of the new structure that was really the issue.

Advertising that the old bridge would only charge twenty cents per trip, rather than the twenty-five cents per trip anticipated for the new bridge (weekly passes were cheaper), Frazer was likely concerned that the old bridge company would lose its revenue. During 1959, the old span recorded a profit of $685,000. The District of North Vancouver pocketed $215,000 of that while the balance was split between the City of North Vancouver, the City of Vancouver and the Dis-

The bridge had been expensive in more ways than one; closure day would be a relief to all. OTTO LANDAUER OF LEONARD FRANK PHOTOS, JEWISH MUSEUM ARCHIVES LF-37085

trict of West Vancouver. Hard money to part with, given that the stipend represented just over 5 percent of the District of North Vancouver's annual budget.

On Monday, April 5, *Vancouver Sun* photographer Deni Eagland was standing on Pier 15 when he snapped a picture of the first of the two 96.5-tonne northside bottom chords as it was lowered into place. The steelwork was now only three months from completion, and the big question on Don Jamieson's mind was whether the two spans would meet as planned. Two months later he was close to having an answer to that question when the two cantilevered arms were separated by only twelve metres. In order to ensure that the bridge was aligned properly, Art Pilon fashioned a slingshot from rebar, which he used to shoot a line over to the south side to assist with the alignment process. Bob Dolphin recalled that they had to perform such tasks on cloudy days, as the sun's heat would distort the unconnected steel at the front end by as much as forty-six centimetres, making any accurate measurement impossible.

Don Jamieson's earlier anticipation that the link would occur sometime between June 8 and 10, weather permitting, was only slightly off given that on June 7 Charlie Geisser laid the 41.1-metre boom of Traveller No. 1 across the void that separated the two spans. Lou Lessard needed no invitation to be the first across the bridge. Stepping gingerly on the boom's cross-braces, he waltzed across the awkward link, noting to the visiting reporters that "There is nothing to be scared of."[283] He later stated, "This bridge has always been a challenge to us. I did it just so I could be the first." It was, of course, his privilege. The excitement was palpable and in the final few hours before closure, a fever caught the men, motivating

them to register a record 192 tonnes of steel erection in just eight hours. "Nobody was driving them," said Jamieson. "They did it because they wanted to."[284]

Closure was one of the most interesting aspects of the whole bridge construction. Four suspended panels extended out from each cantilevered arm, each pinned during construction,

Art Pilon with a slingshot to measure the distance between the north and south cantilevered arms. PHOTO BILL CUNNINGHAM, *THE PROVINCE*

but during closure controlled by eight 565.6-tonne Tangye hydraulic jacks. Two jacks were installed on each of the bottom chords of the south span and one jack on each chord (top and bottom) of the north span. In a typical bridge, the suspended panels are built onshore and hoisted up from barges, but the currents at the Narrows made that option impossible.

Upon completion of the steel erection, the two sides of the suspended span were expected to be separated by no more than about 20 centimetres. To close the steel, the pins would be removed and the hydraulic jacks activated to move each side forward, 12.7 centimetres on the north side and 7.6 centimetres on the south side. The last 30 metres of each side of the suspended span were hinged so that as the two sides were jacked together they could be elevated to facilitate overlap of the lower plates which would then be pinned in pear-shaped slots to hold them fast. Once the hinged spans were lowered back to their resting positions and the topside plates joined, the lower pins were locked into place and the jacks removed. The steel, however, still needed to be able to expand and contract with temperature variations, and in order to accommodate this, telescoping beams had been installed to absorb the steel's travel.

Booming down the steel for one of the final connections. PHOTO BOB DOLPHIN

Norm Atkinson is first across the structural steel of the finished bridge, although Lou Lessard had the privilege of walking the traveller's boom laid over the gap between the two cantilevered arms earlier in the day: "This bridge has always been a challenge to us," he said, "I did it just so I could be the first." PHOTO BILL CUNNINGHAM, *THE PROVINCE*

At 2:15 p.m. on June 7, the first of the last two bottom chords, a 12-metre, 10-tonne member, was lowered into place. As the chord was still dangling in the hooks of the traveller, Norm Atkinson walked confidently across it, the first man to tread the actual steel of the new bridge. John Arnett of the *Vancouver Sun* documented the scene: "'Atta boy, Norm,' one of his fellow workers shouted. And Atkinson's grimy face creased into a wide grin as he called back. 'Bring on the cameras!'"[285]

Following hotly on Atkinson's heels was ironworker Art Street, and after him Don Jamieson and then Carl Stanwick, resident engineer for Swan-Wooster. Both excited engineers reflected the satisfaction of a job well done. In the *Vancouver Sun* the next day, editorial page cartoonist Len Norris drew a caricature of an ironworker referencing Lou Lessard striding proudly across the gap of the incomplete bridge to meet a well-groomed government official on the other side with his hand out. His sign said it all: "Stop, Pay Toll."[286]

Later that same day, "Fearless Phil" (another of his monikers) Gaglardi, added his name to the roster of first-across-the-bridge when, suited up with a life jacket and hardhat, he stepped gingerly across the top chord to tighten one of the last of the 400,000 bolts used in the bridge. As he came over to the other side, he was met with a round of cheers and applause from the thirty ironworkers working that day.

Although the steelwork was largely finished, it would be another eight days before the final adjustments could be made to the suspended span. The weather had to cooperate, and June 15 offered the first opportunity. No technical problems were anticipated as the jacking system had been tested on the completed south half of the suspended span the previous November. In concert, the four south-side jacks had been activated to extend the suspended span a fraction of a centimetre, and in a series of subsequent minuscule budges, the first a mere 3.175 millimetres, a full 5.08 centimetres was achieved by day's end. As the span budged forward at glacier speed under Don Jamieson's watchful eye, copious measurements were taken. This process was now ready to be applied to both spans.

On closure day it was raining, which reduced the temperature variations that were critical to the final adjustments. With microscopic accuracy, for over four hours readings were called out over six telephones and four loud hailers as the spans were jacked ever so carefully together. It was amazing to all that the final adjustments on such a colossal structure would come down to a matter of a fraction of a millimetre, the width of the gap of a spark plug. Planned in minute detail, the locations of all thirty-seven participating engineers and ironworkers

Highways Minister Phil Gaglardi about to tighten one of the last of the 400,000 bolts used to build the bridge. PHOTO P. STANNARD, DOMINION BRIDGE CO. LTD.

was set to paper, much like a screenplay, with each location responsible for carrying out a tiny portion of the whole operation.

Following this delicate operation, the roadbed had to be poured. A special concrete was mixed from saturnalite, a Saturna Island shale that gave the concrete a reddish hue. It was aerated and cooked at 1093.3 degrees Celsius to make it lighter—up to 40 percent—than the usual crushed gravel used to manufacture concrete. The approximate 3,600-cubic-metre pour was applied to a depth of fourteen centimetres over a bed of small steel protrusions called Nelson studs over which was placed a steel mesh, forty-five kilograms of steel per cubic metre of concrete. Once cured, a five-centimetre layer of asphalt was spread over that.

A month after the final connections were made, Tom McGrath launched an attack on his old employer when he asked the BC Labour Relations Board to cancel the certification of the Local for employees of Dominion Bridge. McGrath had been working steadfastly to form a Canadian ironworkers union, the Canadian Ironworkers Union Local 1, since resigning his position from Local 97. He claimed that Dominion Bridge employees were governed more by a company than they were by the union. Sid Stewart, father of Alan Stewart who had died during the collapse, was now the Local's business agent. Stewart wrote a letter to all the Local's associated contractors, asking them for help in defeating the new union that members of Local 97 had voted strongly in favour of establishing at a meeting on May 18, 1960. McGrath's efforts appeared to be paying off as he claimed to have signed up "the great majority" of ironworkers in the province. At the same time, he was applying for certification with four companies in the Steel Erectors Association.

The International was far from pleased with the new development, and according to Colin Glendinning's son Patrick, some ironworkers helping the fledgling union were targeted by the powerful organization. One day Colin, who had been kicked out of the Local for helping organize the new union, received an urgent phone call from Jim English to get out of the house, that he had heard that some men from the International were coming to pay him a visit. Glendinning decided to stay and face them, but sent his wife and children off to the movies just in case there would be violence. Within minutes a heavy boot kicked at the door and a loud, aggressive voice shouted, "Glendinning, come out, we want to talk to you!" Colin opened the door with a twelve-gauge shotgun in hand, and the two men before him stepped back, but threatened him with bodily harm if he continued to help organize the new union.

Although it initially appeared that the Canadian Ironworkers Union would

become a valid competitor to Local 97, over the next decade battles both in the court and on the jobsite finally sucked the life out of the new venture. By 1968 membership had dwindled to only five, but there was still enough juice to sue the International, and in 1972 a court awarded the now-defunct union $30,000 in compensation. After

Lou Lessard holding gap gauges and a file ready for closure. The distance between the two huge cantilevered arms on closure day was that fine.
PHOTO BILL CUNNINGHAM, *THE PROVINCE*

Colin Glendinning and Charlie Geisser, still grinning a year after the collapse.
PHOTO BILL CUNNINGHAM, *THE PROVINCE*

retiring his union's debts, McGrath turned the $2,500 surplus over to another struggling union. Glendinning, recognizing that his career as an ironworker was over given that he was no longer a member of Local 97, moved on to other forms of construction, but the stories that he would tell his children and grandchildren about his time on the steel suggested that he missed the trade. The trade and the bridge had provided him and dozens of other men with stories to last a few lifetimes.

On July 13, Premier Bennett announced that the new span would be open for business on August 25, but that the connecting highway on the North Shore wouldn't be ready for some months. When asked what the new bridge would be called, the premier said, simply, "Just new Second Narrows Bridge."[287] The name of the new bridge, which was hardly creative, nor reflective of the trials experienced to erect it, was as close to that of the old bridge as it could possibly get, even though Minister Gaglardi had promised not to call it that. Murdo Frazer, despite his earlier threats, uttered not a peep.

Opening day, Thursday, August 25, 1960, almost four years and one month since construction began, dawned cloudy with expectations of rain. The temperature at the 3 p.m. ceremony was a modest 14.4 degrees Celsius when survivor Bill Wright snipped the blue ribbon in the drizzle before one thousand onlookers, thus opening the $26 million structure to the public. For three days, travel across the new bridge would be free before the toll-takers would begin to exact their dues. Although uniformed toll employees were in their booths at the south end of the bridge on that first day, they were only there to hand out a controversial souvenir booklet published by the Social Credit League extolling the virtues of the new bridge. Many members of the public responded by raising concerns about the appropriateness of public servants handing out Socred propaganda. The head of the BC Automobile Association, Clarke Simpkins, was in trouble too, because he had handed over his customer list to the Socreds so that they could mail out the pamphlet to all his members.

As cars and trucks whizzed over the new bridge at a heady forty miles per hour, five miles per hour faster than the Lions Gate Bridge, many were wondering where all the traffic was. Were habits that hard to break? For two weeks after opening, commuter volume on the new structure was still low while the Lions Gate Bridge was jammed as usual. In fact, traffic was so light on the new bridge that on the first Sunday of October, a bear cub was seen ambling across the span. Traffic was soon to pick up, however, when the new Upper Levels Highway opened on Saturday, March 4, 1961.

The rating of the new bridge, at completion, was expected to be 800 vehicles

Before the old Second Narrows Bridge was demolished in 1970, three bridges crossed the Second Narrows: the new Second Narrows Bridge, the old rail and car bridge, fondly known as the "Bridge of Sighs," and the new rail bridge. NORTH VANCOUVER MUSEUM AND ARCHIVES

per lane per hour, 300 vehicles per lane per hour less than the engineer's esti-mate for a toll-free structure, and 500 vehicles per lane per hour less than that envisaged by the crossing committee seven years earlier. The real traffic flows were unknown, because the government refused to release its records to the press. But the week before the Upper Levels Highway opened, an informal *Vancouver Sun* survey recorded 2,000 southbound cars per hour (averaging 667 per lane) cross-ing the new bridge at peak times, increasing to 2,300 (767 per lane) after the opening of the highway. Toll officials would only offer the vague comment that they had noticed a significant increase during peak times. Volumes would also get another boost when the Burnaby-through-Coquitlam connection to the Port Mann Bridge opened, but that was still some months away.

A year and a half later, municipal traffic engineers set out to determine how many vehicles were crossing where, and what they discovered shocked some and caused others to say, *I told you so*. Out of six city bridges, the Second Narrows Bridge was dead last in traffic, at 17,000 crossings in a twenty-four-hour period. But according to the 1954 crossing committee report, which forecast that the Second Narrows Bridge volume would increase from one third that of the Lions Gate Bridge to one half by 1976, traffic volume on the new structure was way ahead of schedule. It was already registering half of the estimated 33,000 crossings recorded at the Lions Gate Bridge for the same twenty-four-hour period. Traffic volume was obviously growing at the Second Narrows, something that Gaglardi said would happen, increasing from about 12,000 cars in a twenty-four-hour pe-riod at opening to 17,000 one and a half years later. It was quickly proving itself to be an indispensable link in the Lower Mainland's transportation network.

With such low traffic volumes it seemed that it might take forever to pay for the structure, but toll booths had become a fixture on BC bridges, and the travel-ling public, like it or lump it, were getting used to them. So, it was a big surprise when the premier announced on April 18, 1962, that plans were afoot to lift the tolls on provincially controlled bridges. Some were saying that it was about time given that the Lions Gate Bridge had already been paid for again and again. It would take almost another year, however, for Bennett to get his ducks in a row, but on April 1, 1963, tolls ended on the Second Narrows and Lions Gate bridges as well as bridges at Kelowna, Nelson and on the Agassiz-Rosedale Bridge cross-ing the Fraser River at Hope. The $32 million required to cover the bridge debt had been extracted from that year's budget surplus.

The government largess, however, spelled doom for the old municipal bridge next door, which couldn't afford to lift its tolls and was therefore forced to close to vehicle traffic. It was a sad end to a structure that had once been the only fixed link between Vancouver and the North Shore, but it was also no secret that the bridge had caused its owners a plague of ills. Not only did it trigger a major row between Vancouver and Ottawa, but it also managed to accumulate a million dollars in repair bills as well as cause the bankruptcy of the North Shore municipalities. The bridge, which was eventually sold to the CNR for one dollar, continued to serve rail traffic until it was replaced by a new rail bridge whose construction began mid-1966 at a cost of $8 million.

In a photo taken in 1969 all three bridges were still in place across the Second Narrows. Despite the rail bridge being younger, it looks antiquated beside its still-modern, but imposing, six-lane-deck-bridge neighbour—a neighbour with a costly history. Two days before the bridge's opening ceremony, a small group recognized this cost by gathering at the southeast end of the bridge to unveil a modest bronze plaque set in a concrete plinth erected by the Ironworkers District Council of Western Canada. The plaque read: "In memory of those who lost their lives in process of construction of this, the Second Narrows Bridge." Most notable among the guests was not the premier, nor Highways Minister Gaglardi, but the widow of John A. Wright, who had died during the collapse. Mrs. Wright had been offered the option of either snipping the ribbon of the new bridge or unveiling the plaque. She couldn't bear the thought of cutting the ribbon on a bridge that had taken her husband's life, but after thinking it over decided that it would be an honour to unveil the plaque that listed her late husband's name with the names of the other men killed on the bridge.

However, time not only has a way of healing, but also of forgetting. Tom McGrath's son Kevin, also an ironworker, was a Local 97 business agent in the early nineties. When he saw a resolution on a BC Federation of Labour convention agenda to lobby the government to have the bridge renamed the Ironworkers Memorial Second Narrows Crossing, he needed no encouragement to support it. He followed up by lobbying politician after politician to educate them and gain their support. Kevin remembered being surprised to encounter some resistance:

> I found it actually pretty sad for elected people to think like that, but anyway, despite all that I just kept going... and I wrote a letter to the provincial government, and the Highways minister at the time was Jackie Pement, and she called up and said, "Let's have a meeting and talk about this"... and she decided, "Yeah, I'm the Highways minister, and I think there is a good reason and good logic to change the name of the bridge," and she went ahead with it and the government stood behind that and they did change it.

Unfortunately Reverend Dr. George Turpin, the Shaughnessy Hospital chaplain who had presided over the memorial thirty-six years earlier, and who foresaw that "When the span is completed, it will be a giant memorial to the loved ones who are gone,"[288] would not live long enough to see his tribute officially become a reality. On a brilliant Friday, June 17, 1994, Premier Mike Harcourt, surrounded by the remaining survivors, ceremoniously unveiled the plaque that dedicated the bridge its special status. The Ironworkers Memorial Second Narrows Crossing has not only become a permanent public monument to the men who lost their lives building it, but it is also a reminder of the hazards faced daily by the men and women "who strive so brilliantly and dangerously to give us a modern world."[289]

Epilogue

Of the seventy-nine men working on the bridge that tragic day in June, I only managed to track down and speak to a few of them. Many had already passed on, some were reluctant to speak to me and others had relocated to parts unknown. Age being the ultimate deceit, I wondered how clear their memories would be. I needn't have worried. Although some details were fuzzy, it was almost as if the tragedy had occurred yesterday for most of these men.

The day had not only moulded who they had become, but for a select few it had made them into minor celebrities. For others though, especially the widows and their children, it was a different story, and in my naiveté I was unprepared for the emotion that still wracks some of these families. The transcendence of their grief through the decades and generations is a powerful testament to their loss, a loss that was never so apparent than when Paul McDonald reverently opened a small box to show me the contents of his father Murray's pockets on the day of the collapse, now coated with a thin malachite patina, or when the late Maureen Colley related the story of an unfinished sweater in her sister Diane's house with a bridge design on it that she and her late father, Joe Chrusch, were knitting together.

Each house I entered had pictures and memorabilia, some on display, some hidden away as if to forget. For Norm Atkinson it was a proud few framed photos on his wall and his ironworker's well-worn tool belt; for Gary Poirier, who ushered me into his little room on Hastings Street in downtown Vancouver to look at his collection of bridge photos and posters that walked across his walls, it was like living the steel that he had tread decades before. For Patrick Glendinning, Colin's son, it was a shattered lifejacket that had saved his late father's life, still grimy with the grease of a working man.

Although you can see faces of despair at the ceremony every June 17, you can also feel the hope and comfort that settles over this rough congregation that gathers below the memorial garden at the southeast corner of the bridge just behind the cement plinth that bears the men's names. That the ceremony has a special home is only due to the generosity of Cliff and Sharon Nordquist who permit the use of their beautiful garden every year for the service. It usually rains, or at least it has the last three ceremonies I have attended—a stark contrast with the actual weather on that fateful day fifty years ago. This, I suppose, is fitting.

When Reverend Barry Morris finishes his prayer every year and the piper skirls his lament from the peak of the garden, slowly leading the procession down the short hill to the memorial, it all comes together in the simple gesture of a placed wreath, a bowed head, a wiped tear and a doffed hat. Over the years the memorial-day demographics have changed. Gone are many of the survivors, but loyally replacing them are their children, the children of the men who perished, other relatives and even the next generation—some of whom are ironworkers themselves. Union families not only believe in solidarity, but in tradition.

If you listen closely to those around you, you will hear stories of the legacies that drive these families. Some have lost more than one family member to the trade, but resolutely, sons, brothers, fathers and now even mothers and sisters still continue to walk the iron. Risk and injury still dog them, almost as if it is an inherent price of admission. Donny Geisser remembered answering a knock at his door one evening just after the bridge was finished. There stood big John Olynyk with a vacuum beside him on the porch. After buying one, he listened to Olynyk's story. During reconstruction of the bridge he had stepped between two barges and ended up in the ocean again. Right then and there, he had decided that he had cheated death once too often.

Bridge engineers, too, face the same perils. Peter Buckland, partner of the respected Vancouver bridge engineering firm Buckland and Taylor, recounted a recent tragic story. Anthony Freeman, son of Sir Ralph Freeman who gave Peter his first job, died from injuries sustained during a falsework collapse beneath the Vasco da Gama Bridge at Lisbon, Portugal. Although the accident occurred on April 10, 1997, his injuries didn't steal him from this world until July the following year, after he lay in a coma all that time. His father died shortly after—they say of heartbreak. It is a grim irony that the father was the expert witness called to give evidence into a falsework collapse thirty-nine years earlier.

Freeman had been right when he wrote to Commissioner Lett that Dominion Bridge would be embarrassed by the collapse. Despite publishing a comprehensive history/inventory of its most prominent Canadian achievements, DB excluded the Second Narrows Bridge from its *Cavalcade of Steel* (DB, 1972) and *Visions of Steel* (AMCA International, 1982). The company record for the former publication includes only Dorchester Boulevard in Montreal for 1960, the year the bridge opened, and in the latter publication that year is totally vacant. Bob Harris, a DB erection engineer, wrote that "when DB closed its Vancouver operation, c. 1975, the files and drawings were thrown out of the window in a great talus slope,"[290] thus ensuring that these historic documents were permanently beyond review and perhaps, criticism.

Not a great statement, or legacy, to the men who died in the employ of the giant steel erector. This neglect is one of the reasons why it is important that we remember these men by calling the bridge by its correct name, the Ironworkers Memorial Second Narrows Crossing. Many people still call it the Second Narrows Bridge because it is shorter and has become habit, but in the last decade or so, an abbreviated "Ironworkers" or "Ironworkers Memorial," has gained popularity, thanks in part to conscripts like CBC Radio *Early Edition* host Rick Cluff and his colleagues.

Whoever travels daily over either of the two Burrard Inlet bridges knows that what Hal Denton, spokesman for the Seymour Ratepayers Association, said in 1954 is as relevant today as it was then: "those of us living in the areas most affected feel a start should be made right away on it . . . the traffic situation right now is beyond a joke . . . we think it's about time someone started the ball rolling on this."[291] At that time, Hal was talking about a new crossing either at the First or Second Narrows. In fact, Minister Gaglardi reserved 11.3 hectares just east of the Lions Gate Bridge in 1956 for another First Narrows crossing that of course never materialized. A third crossing is now well overdue, given that the population of the city has more than doubled since the Second Narrows Bridge opened in 1960.

This argument did briefly die away after the opening of the new Second

Narrows Bridge, but not for long. In 1967 there was a proposal for an Indian Arm Causeway to be built east of the new bridge at a cost of about $28 million. About the same time, the federal government began investigating another crossing as well when public acrimony stalemated municipal initiatives. One option was a stunning cable-stayed structure across the First Narrows designed by Peter Buckland while working for Swan-Wooster–CBA (joint venture) that would have had enough capacity to retire the Lions Gate Bridge. However, plans were shelved several years later when anti-freeway activists kicked up a storm and the NDP government reallocated the $27 million it had set aside for the crossing to research and development of a rapid transit system for Vancouver.

The most recent study, in November 1993, looked at nine crossing options, including widening of the existing three lanes on the Lions Gate Bridge; modifying the existing lanes on that bridge to four; building a new four-lane First Narrows structure; building a new five-lane First Narrows structure; tunnelling from the Upper Levels Highway to Georgia Street; boring a shallow tunnel from Capilano Road in North Vancouver to Stanley Park; building a combined tunnel and causeway near Brockton Point; building a tunnel mid-harbour; and finally, widening the Ironworkers Memorial Bridge. The Province acted on the first and cheapest option, a delicate aerial act performed by Buckland and Taylor that has not only made the bridge safer but lengthened its life. This rejuvenation, however, did little to address the city's growing traffic issues.

Perhaps we all need to heed the advice of Anthony Downs, a senior fellow at the Brookings Institute think tank in Washington, DC, who said that "The idea of getting rid of congestion altogether is a delusion. Traffic congestion is an inescapable result of the way we organize society. It's not going to go away—in fact, it's going to get worse."[292] Recognizing that we will probably never catch up to traffic demands is frustrating but nevertheless a reality.

Finally, I could not comfortably finish this book without some sort of personal comment about the collapse. Although it is an irrefutable fact that McKibbin and McDonald were responsible for the dimensional errors on the critical falsework design sheet, to conclude that these errors were the *entire reason* for the collapse would be wrong. In addition to the company's unwritten policy of isolating major projects, thus removing them from the mainstream checking process, of burdening McDonald with an impossible load and placing an important design feature in the hands of a relatively inexperienced engineer, there was the startling admission by Professor Hrennikoff that the upper grillage I-beams were weak with respect to buckling strength.

And then there was Professor Armstrong's investigation of beam no. 2, and his discovery that it was from a different heat and parts of it had a lower yield point and tensile strength than specifications required. That falseworks were considered pieces of equipment and therefore not subject to the same scrutiny given steel erection, was also problematic, as was the difference in the shear strength calculation formulas between CSA and AASHO, the former of which was 30 percent more tolerant and according to Professor Hrennikoff, "at some level, actually unsafe."

The royal commission too, left some unanswered questions. For instance, why was Robert Eadie, vice-president and manager of the company's Eastern Division, called to testify rather than the local and more knowledgeable individuals who were building the bridge? And why wouldn't the company let Angus McLachlan on the stand? What did he know that the company didn't want made public? The comment from his son that he would have "told all" is revelatory.

I also have a feeling that the company knew something about who the mystery man was who partially corrected one of the errors. Robert Eadie testified that "nobody knows definitely who made that note in pencil," but then stated that the executives knew it was made prior to the accident. If they knew that, they must have known something more. Not that blame was the intent of the royal commission, but it would have been instructive to know if it was a breakdown in processes that caused the error not being acted upon or reported. And why did John Farris not apply his copious probing skills more thoroughly to Eadie's testimony? When Eadie said that the company didn't know "definitely" who made the change, Farris responded, "Before I pursue that . . ." but never did return to the question.

With no further explanation, some might conclude that it was one of the two engineers who made the change, but then we go to the number of character descriptions of both McKibbin and McDonald, which in my opinion makes that theory unlikely. McKibbin's letters home talked about how lax bridge building was in Canada, and how, if he was to cause the death of a workman, he would walk off the job immediately. Given McKibbin's upbringing, his devotion to his career, his earnestness and the fact that he passed the BC Engineering qualifications with "very high recommendations," it would be hard to imagine that he could have discovered his own mistake and then have been so cavalier as to not perform the rest of the calculation or tell McDonald about it. Wouldn't he have at least wanted to know how the change would have altered the load-bearing capacity of the beams? I think so.

That then swings the onus onto McDonald as being the author of that pencilled note, but despite repeated attempts during the royal commission to find fault with his character, all failed. Lieutenant Colonel Harry H. Minshall said it best when he stated that he had worked with McDonald and knew him to be a thorough and conscientious individual who, had he even had the slightest inkling that the falsework was faulty, would have stopped the erection immediately. And Eadie stated that McDonald "was an engineer with the second-most experience in erection problems and we considered him a very competent engineer."[293] Does this sound like a man who would find a critical error in a design sheet and then not act on it? Again, I think not.

McDonald and McKibbin's role in the collapse, though tragic, was only one factor in a complex series of inadvisable engineering practices, complacency with respect to falseworks, questionable steel quality, structurally imperfect beams and an unsafe (at some level) columnar formula that all conspired against the N4 falsework. As Minshall said, the collapse was truly "accidental to human knowledge."

Acknowledgements

There is a saying that it takes a village to raise a child. Well, it is much the same for writing and publishing a book. Although the author receives the recognition, it would be an injustice to suggest that the process is the result of a singular achievement. There would be no book at all were it not for the cooperation of the dozens of people that I interviewed, most prominently the remaining few survivors of the collapse: Lou Lessard (thanks, Lou, for reading the manuscript and for your advice); Norm Atkinson; Jim English; and Gary Poirier. Unfortunately, the late Bill Stroud, the survivor's unofficial spokesman, was too ill to be interviewed, but I listened intently as his widow, Donna Stroud, and his buddies, reverently told me stories about him. Thank you for that. Thank you also to friend, historian Robin Inglis, for suggesting that this story was a gap in the community's history record. I must also acknowledge Peter Buckland, of the international bridge engineering firm, Buckland and Taylor, who has written a generous foreword (thanks for reading the text, Peter, and for your advice).

A big thank you also goes to Peggy Stewart who gave me the first names on my path to investigate the tragedy. Those names lead me to others to whom I owe a debt of gratitude. The story would not be as complete without their input. They are, alphabetically, as follows: the late Marika Ainley, author, whose advice was so very important; Doreen Armitage, author (thanks, Doreen, for your advice); Dave Beatty, former Local 97 president; Tom Berger (OC, OBC), jurist, politician (thanks, Tom, for your review of and advice on chapter eleven); Jim Bisset, witness; Ralph Bower, retired *Vancouver Sun* photographer; the late Jack Bridge, VFD rescue diver; Kim D. Carpenter, granddaughter of the late Otmer C. "High" Carpenter; Charlene Chang, VGH archivist; Diane Chrusch and the late Maureen Colley, the late Joe Chrusch's children; City of Vancouver Archives staff; Cecil Damery, Local 97 president; Jean Dierks, daughter of the late Gordon MacLean; Hugh Dobbie, former steel detailer; Robert Dolphin, P. Eng., retired (thanks, Bob, for reading the manuscript and for your advice); Sandra Eadie, granddaughter of the late Robert S. Eadie; Bob Gaglardi, son of the late Highways Minister Phil Gaglardi; Allan Galambos, bridge maintenance; Phil Gamble, retired op. eng.; Gary Geddes, author of *Falsework*, a poetry book about the collapse (thanks, Gary, for your support and your advice); Don Geisser, retired ironworker, rescue diver and son of the late Charlie "Grumpy" Geisser; Chrysta Gejdos, one of three authors of the film script, *19 Scarlet Roses*; Fran Glendinning, spouse (at the time of the collapse) of the late Colin Glendinning, and Patrick Glendinning, son of Colin; Government of Canada Archives staff; Ann Gregoire, daughter of the late Murray McDonald; Tudy Grelish, retired Dominion Bridge contract supervisor; Charlie Guttman, son of the late Eric Guttman; Bill Hadley, retired VFD rescue diver; Richard Hallaway, retired NVFD fireman; Eileen Haynes, retired VGH nurse; Ruth Herman, niece of the late Norm Eddison; Don Heron, retired ironworker; George Hoffman, retired carpenter; Ron Hyrnuik, general manager, Bridge Studios; Jean Jamieson, spouse of the late Don Jamieson; Ralph Johnson, retired carpenter; Janine Johnston, archivist, Jewish Community Centre; Allan A. Kay, P. Eng.; Graeme Kelleher, engineer, friend of the late John McKibbin; Julia Kim, WorkSafeBC; Carol Lasko, daughter of the late Bill Lasko; James Leland, Local 97 business agent; Paul McDonald, son of the late Murray McDonald (thanks, Paul, for your friendship, support and the many stories about your dad); Kevin McGrath, son of the late Tom McGrath, ironworker and bridge-renaming advocate; Bill and Rob McLachlan, P. Engs., son and grandson respectively of the late Angus McLachlan; Peter McRuvie, nephew of the late John McKibbin (thanks, Peter, for reviewing chapter four and for providing John's and Sam McKibbin's letters, without which the text would not be as rich); Byron Maine, retired Dominion Bridge paint inspector; David Milne, a.k.a. "Davy Crockett," retired ironworker; Peter Mitchell, P. Eng., iron ring ceremony; Jim Morrison, sound engineer; Barbara Murray, widow of John McKibbin (thanks, Barbara, for your courage and your generosity of spirit); Cliff and Sharon Nordquist, who annually permit their wonderful garden to be used for the memorial service; North Vancouver Museum & Archives staff: Daien Ide, reference historian; Janet Turner, archivist; and Francis Mansbridge, former archivist; Phil Nuytten, rescue diver; Art Pilon, retired ironworker and rescuer; Jim "Skip" Pratt, retired Dominion Bridge construction administrator and rescuer; the late John Prescott, P. Eng.; Colin Preston, CBC archivist; Andrew Priatel, owner of the family-run Lynnwood Inn; Province of British Columbia Archives staff; Andrew Roberts, retired VPD diver; Chuck Robinson, witness; Red Robinson, media personality (thanks, Red, for the two recordings); Mel Rothenberger, author of *Friend O' Mine, the Story of Flyin' Phil Gaglardi*; Bob Sexsmith, P. Eng., retired UBC professor; George Shephard, P. Eng., retired; Don Sinclaire, retired military rescue specialist; Denny Sinnot and Neil Sutherland, Tidal Records, Victoria; Frank Smirfitt, retired draughtsman; Al Snider, retired ironworker; Charles Todd, retired NVFD fireman; UBC Rare Books and Special Collections staff; Vancouver Public Library staff; and Dr. Don Warner, attending physician.

I would also like to extend my gratitude to Harbour Publishing for offering to publish this story as well as for its commitment to regional history. Special credit is due to Vici Johnstone, general manager; Silas White, editor; Anna Comfort, production manager; Erin Schopfer, Greg Virc and Teresa Karbashewski, editorial/production assistants; and Marisa Alps and Rachel Page, marketing, among others.

Friends—too many to mention—as well, offered advice, support, encouragement and stories. For that I am grateful. Finally, I would be remiss in not extending a profound debt of gratitude to my wife, Joan, who continued to reassure me that there was a story in all the interviews I was conducting. My children too offered unconditional support: my daughter, Amanda; and my son Ian and his wife, Carlie. I would also like to acknowledge my father, John V. Jamieson, for his support and his memory of that tragic day.

If I have missed thanking anyone in these pages it is an oversight for which I apologize. Although several experts read the text for its correctness and authenticity, any errors or omissions are mine alone. All reasonable efforts were made to locate holders of copyright and the author would like to know if any further acknowledgements are necessary.

Notes

1. Mel Rothenburger, *Friend o' Mine: The Story of Flyin' Phil Gaglardi* (Victoria: Orca, 1991), 111–112.

2. Tourist and Highways Conference, May 6, 1955.

3. *Vancouver Daily Province* (North Shore Bureau), Nov. 19, 1951.

4. Jack Wasserman, *Vancouver Sun*, Apr. 16, 1953.

5. Uncredited newspaper clipping (Mar. 27, 1954) from North Vancouver Museum & Archives.

6. *Vancouver Daily Province*, Sep. 29, 1954, 21.

7. Tom Hazlitt, *Vancouver Daily Province*, Jun. 16, 1954.

8. *Vancouver Province*, Jan. 11, 1955.

9. *A Report on Burrard Inlet Crossings*, Committee on Burrard Inlet Crossings, Nov. 1954, 10.

10. Gordon McCallum, *Vancouver Province*, Jan. 20, 1955, 1.

11. *Vancouver Sun*, Mar. 29, 1955, 17.

12. *Vancouver Province*, Apr. 1, 1955, 46.

13. *Vancouver Sun*, March 29, 1955, 17.

14. *Vancouver Daily Province*, Jun. 1, 1940.

15. *Vancouver Province*, Jun. 1, 1955.

16. Uncredited newspaper clipping (May 7, 1955) from North Vancouver Museum & Archives.

17. *Vancouver Sun*, May 10, 1956, 3.

18. *Vancouver Sun*, Feb. 27, 1957, 20.

19. Debbie Smith. "Second Narrows Bridge Information Manual" prepared for Regional Highways Engineer, Ministry of Transportation and Highways, Region I: Burnaby.

20. *Vancouver Sun*, Feb. 9, 1956, 1.

21. *Vancouver Sun*, Feb. 10, 1956, 27.

22. *Vancouver Sun*, Feb. 14, 1956, 38.

23. *Vancouver Province*, Feb. 25, 1956, 5.

24. *Vancouver Herald*, Jul. 13, 1956, 3.

25. Hal Dornan, *Vancouver Province*, Jul. 17, 1956.

26. *Vancouver Province*, Jul. 20, 1956.

27. *Vancouver Province*, Aug. 6, 1957, 17.

28. Alex MacGillivray, *Vancouver Sun*, Jul. 6, 1957.

29. "Field Report," *Construction World*, Apr. 1958, 29.

30. In Jun. 2007, the site was returned to the private sector through a $40 million sale to Larco Investments Ltd., a company controlled by West Vancouver's Lalji family.

31. Interview with Jim "Skip" Pratt.

32. "Field Report," *Construction World*, Apr. 1958, 28–29.

33. Ibid.

34. John McKibbin's letters home, compiled by Peter McRuvie (nephew).

35. Deposition of R.S. Eadie, Royal Commission Proceedings, 681.

36. DB memo from Murray McDonald dated Nov. 25, 1957.

37. John McKibbin's letters home, May 28, 1957, as compiled by Peter McRuvie.

38. DB memo from Murray McDonald dated Nov. 25, 1957.

39. Interview with Jim 'Skip" Pratt.

40. DB memo to Murray McDonald dated Nov. 25, 1957.

41. John McKibbin's letters home, compiled by Peter McRuvie.

42. An article in the *Canadian Journal of Civil Engineering* brings the actual fatality rate into perspective. Stats gathered by D.E. Allen for his 1992 submission (Vol. 9) report that the annual death rate per million people for collapses during construction was 20, negligible when you consider that 360 people per million died in automobile accidents, 500 from smoking, 550 from accidents and 726 from disease.

43. John Prescott, Royal Commission, 913.

44. Conversation constructed referencing George Gareth's 1998 CBC interview with Norm Atkinson, Bill Stroud and Lou Lessard.

45. Paul Luke, *Vancouver Province*, Nov. 2, 2003, A47.

46. *Shorter Oxford English Dictionary* (2nd Ed.), Vol. 2, 2794.

47. Ralph Freeman, Royal Commission, 544.

48. A note at the bottom of John McKibbin's calculation sheet stated, "Use 4 WF 160 Without Stiffening."

49. John L. Farris cross-examining R.S. Eadie, Royal Commission, 685.

50. R.S. Eadie, Royal Commission, 693–694. / Contract No. 2 between the British Columbia Toll Highways and Bridge Authority and the Dominion Bridge Company Ltd., 19, paragraph 2–2–3.

51. John Prescott interview, Jan. 21, 2005.

52. *North Shore News*, May 8, 1958.

53. Dream as originally told to Chrysta Gejdos (BCIT) for the film *19 Scarlet Roses*.

54. Lou Lessard interview by Chrysta Gejdos for the film *19 Scarlet Roses*.

55. Conversation partially constructed from Royal Commission, 230.

56. Conversation partially constructed from Royal Commission, 223.

57. Conversation partially constructed from Royal Commission, 113.

58. Norm Atkinson interview by Chrysta Gejdos for the film *19 Scarlet Roses*.

59. Ibid.

60. George Gareth CBC interview, 1998.

61. Chad Skelton, *Vancouver Sun*, Mar. 16, 2006.

62. Frank Rasky, *Liberty Magazine*, Dec. 1959, 54.

63. Ibid.

64. Gary Poirier interview by Chrysta Gejdos for the film *19 Scarlet Roses*.

65. Ibid.

66. "Bridge Disaster Recalled by Keith Lowe," *North Shore News*, Jun. 26, 2000.

67. Bruce McLean, *The Columbian*, Jun. 18, 1958.

68. Lou Lessard at Jun. 17, 2007 memorial ceremony.

69. *The Span* (Dominion Bridge Co. Ltd., and Subsidiary Companies), Jul.–Aug. 1958.

70. *Daily Mirror*, Sydney, Australia, Jun. 19, 1958, 12.

71. Letter from Sam McKibbin to Barbara McKibbin, Oct. 14, 1958.

72. Letter from Sam McKibbin to *Vancouver Sun*, Jul. 17, 1958.

73. Glen McDonald with John Kirkwood. *How Come I'm Dead?* (Hancock, 1985), 109.

74. Frank Rasky, *Liberty Magazine*, Dec. 1959, 60–61.

75. *Vancouver Province*, Jun. 18, 1958.

76. Jean Howarth, *Vancouver Province*, Jun. 18, 1958, 3.

77. Ed Cosgrove, *The British Columbian*.

78. Interview with Ed Cosgrove: Glen McDonald with John Kirkwood, *How Come I'm Dead?* (Hancock, 1985), 113.

79. *The British Columbian*, Jun. 18, 1958, 6.

80. *Vancouver Province*, Jun. 18, 1958.

81. George Hoffman interview by Chrysta Gejdos for the film *19 Scarlet Roses*.

82. *Vancouver Sun*, Jun. 18, 1958.

83. Ibid.

84. Ibid.

85. Ibid.

86. Ed Cosgrove, *The Columbian*, Jun. 18, 1958.

87. Frank Rasky, *Liberty Magazine*, Dec. 1959, 62.

88. *Vancouver Sun*, Jun. 18, 1958.

89. Ibid.

90. *Victoria Times*, Jun. 18, 1958.

91. Bill Fletcher, *Vancouver Sun*, Jun. 18, 1958, D1.

92. *Vancouver Sun*, Jun. 18, 1958.

93. *Victoria Times*, Jun. 18, 1958.

94. *Victoria Colonist*, Jun. 18, 1958.

95. *Kamloops Sentinel*, Jun. 18, 1958.

96. *Victoria Colonist*, Jun. 18, 1958.

97. Ibid.

98. Ibid.

99. Simma Holt, *Vancouver Sun*, Jun. 18, 1958.

100. *Vancouver Province*, Jul. 7, 1958.

101. Glen McDonald with John Kirkwood, *How Come I'm Dead?* (Hancock, 1985), 109.

102. Ibid., prologue.

103. Ibid., 115.

104. Ibid., 112.

105. Ibid., 114.

106. Ibid., 118.

107. Ibid.

108. *Vancouver Sun*, Jun. 24, 1958.

109. Glen McDonald with John Kirkwood, *How Come I'm Dead?* (Hancock, 1985), 119.

110. *Prince George Citizen*, Jun. 25, 1958.

111. *Vancouver Sun*, Jun. 19, 1958.

112. Reginald H. Roy, *Sherwood Lett: His life and times* (UBC Alumni Association), 158.

113. Ibid., 161.

114. Sherwood Lett, Royal Commission diary, Jun. 17, 1958.

115. Royal Commission, Vol. 1.

116. Sherwood Lett, diary, Jun. 18, 1958, 1.

117. Ibid.

118. Uncredited newspaper clipping (Jun. 20, 1958) from North Vancouver Museum & Archives.

119. Harry H. Minshall, "In Memoriam, Murray MacDonald, Professional Engineer," *The BC Professional Engineer*, Jul. 1958, 21.

120. *Vancouver Sun*, Jul. 17, 1958.

121. Uncredited newspaper clipping (Jun. 20, 1958) from North Vancouver Museum & Archives.

122. *Vancouver Sun*, Jun. 26, 1958.

123. *Vancouver Province*, Jul. 10, 1958.

124. Ibid.

125. *Vancouver Province*, Jul. 19, 1958.

126. Ibid.

127. Royal Commission, Vol. 2, 23.

128. Ibid., 156.

129. Royal Commission, Vol. 1, 149.

130. Ibid., 101.

131. Ibid., 113.

132. Royal Commission, Vol. 3, 219–220.

133. Ibid., 232.

134. Royal Commission, Vol. 3, 289–290.

135. Royal Commission, Vol. 4, 339.

136. Ralph Freeman letter to Sherwood Lett, Aug. 21, 1958.

137. Ibid.

138. Sherwood Lett, diary, 15.

139. *Vancouver Sun*, Sep. 3, 1958, 10.

140. Sherwood Lett, diary, 22.

141. Sherwood Lett, diary, 22.

142. Farris, Stultz, Bull & Farris letter to Jestley, Morrison, Eckardt, Ainsworth & Henson, Sep. 17, 1958.

143. Sherwood Lett, diary, 26.

144. Royal Commission, Oct. 2, 1958, 652.

145. Royal Commission, Vol. 5, 366–367.

146. Ibid., Schedule 12: G.S. Eldridge & Co. Ltd., 11, 14.

147. Ibid., 377.

148. Ibid., 382.

149. Ibid., 383.

150. Ibid., 391.

151. Ibid., 392.

152. Ibid., Appendix 12 (d): Professor Hrennikoff, 8.

153. Ibid., 19.

154. Journal of the Structural Division Proceedings of the American Society of Civil Engineers, *Lessons of Collapse of Vancouver* 2nd Narrows Bridge, A. Hrennikoff, 7.

155. *Vancouver Province*, Oct. 2, 1958.

156. Royal Commission, 500.

157. *Vancouver Province*, Oct. 1, 1958.

158. Royal Commission, 512–513.

159. Ibid., 581–582.

160. *Victoria Times*, Oct. 2, 1958.

161. Royal Commission: F.M. Masters, 633–634.

162. H.H. Minshall, Royal Commission, 644.

163. Ibid., 645.

164. Ibid., 648–649.

165. Ibid., 653.

166. Royal Commission: R.S. Eadie, 673–674.

167. Ibid., 677.

168. Ibid., 680–682.

169. Ibid., 687.

170. Ibid., 690–695.

171. Royal Commission, W.G. Swan, 722.

172. Ibid., 735–736.

173. Ibid., 745.

174. Ibid., 753.

175. Ibid., 757.

176. Ibid., 762–767.

177. J.R. Giese, Royal Commission, 772.

178. Ralph Freeman, Royal Commission, 775–776.

179. Uncredited newspaper clipping, (Oct. 2, 1958) from North Vancouver Museum & Archives.

180. *Victoria Times*, Oct. 2, 1958.

181. Sam McKibbin letter to Barbara McKibbin, Oct. 14, 1958.

182. Sherwood Lett, diary, 27.

183. *Vancouver Sun*, Oct. 7, 1958, 6.

184. Royal Commission: Wm. C. McKenzie, 885.

185. Executive Placement Services letter to Murray McDonald, Jun. 6, 1958.

186. Mark Ketchum, Bridge Aerodynamics website, retrieved Jul. 4, 2008 from http://www.ketchum.org/wind.html

187. Charles E. Andrew, Royal Commission, 893.

188. William M. Armstrong, Royal Commission, 917.

189. Ralph Freeman, Royal Commission, 923.

190. Ibid., 925.

191. John L. Farris, Royal Commission, 933.

192. Ibid., 933–934.

193. Ibid., 938–939.

194. Ibid., 943.

195. Ibid., 944–945.

196. Ibid., 957.

197. Herb Macaulay, Royal Commission, 959.

198. Ralph Sullivan, Royal Commission, 963.

199. Royal Commission, H. Lyle Jestley, 966–973.

200. Ibid., 975–976.

201. Ibid., 980.

202. Ibid., 984.

203. *Vancouver Sun*, Oct. 17, 1958.

204. George H. Steer, Royal Commission, 990–991.

205. Ibid., 999.

206. Ibid., 1011–1012.

207. Norm Eddison, Royal Commission, 1016.

208. John L. Farris, Royal Commission, 1018–1019.

209. Sherwood Lett, diary, 34.

210. DB partnered with the Canadian Bridge Company to form the St. Lawrence Bridge Company to finish the Quebec Bridge, which had collapsed under different contractors in 1907 killing over seventy men. In 1916 the bridge collapsed again under the DB/CBC partnership. The bridge was finally completed in 1918.

211. Royal Commission, Vol. 1, 6.

212. Royal Commission, Vol. 1, 8.

213. Ibid.

214. Royal Commission, Vol. 1, 11.

215. Ibid., 11–12.

216. *Vancouver Province*, Dec. 3, 1958, 1.

217. *The Columbian*, Dec. 3, 1958.

218. *Liberty Magazine*, Dec. 1959, 57.

219. *The Columbian*, Nov. 21, 1958.

220. Eric Lindsay, *Vancouver Province*, Jun. 12, 1959.

221. Denny Boyd, *Vancouver Sun*, Dec. 19, 1958.

222. F.T. Brown letter to Swan, Wooster & Partners, Jan. 19, 1959.

223. Paddy Sherman, *Vancouver Province*, Apr. 23, 1959.

224. *Vancouver Sun*, May 19, 1959.

225. *Vancouver Sun*, Jun. 13, 1959.

226. Simma Holt, *Vancouver Sun*, Jun. 17, 1959.

227. *Vancouver Province*, Jun. 20, 1959.

228. David J. Mitchell, *WAC: Bennett and the rise of British Columbia* (Vancouver: Douglas & McIntyre, 1983), 266.

229. Ibid, 268.

230. *Vancouver Province*, Jun. 23, 1959.

231. *Trade Union Act v. Local 97*, Court of Appeal, Jun. 18, 1959, 7.

232. Ibid., Jun. 22, 1959, 13–14.

233. Ibid., Jun. 26, 1959, 2–3.

234. Ibid., Jun. 24, 1959, 18.

235. Ibid., Jun. 17, 1959, 24.

236. Doug Collins, *Vancouver Sun*, Jun. 24, 1959.

237. *The Columbian*, Jun. 24, 1959.

238. *Vancouver Sun*, Jun. 25, 1959.

239. Thomas R. Berger, *One Man's Justice* (Vancouver: Douglas & McIntyre, 2002), 38.

240. *Trade Union Act v. Local 97*, 60.

241. Ibid., 62.

242. Berger, *One Man's Justice*, 41.

243. *Trade Union Act v. Local 97*, 91.

244. Berger, *One Man's Justice*, 43.

245. *Trade Union Act v. Local 97*, 103.

246. *Vancouver Sun*, Jul. 2, 1959.

247. Doug Collins, *Vancouver Sun*, Jun. 27, 1959.

248. *Vancouver Sun*, Jun. 29, 1959.

249. *Vancouver Province*, Jun. 27, 1959.

250. *The Columbian*, Jun. 29, 1959.

251. *Victoria Colonist*, Jun. 30, 1959.

252. Berger, *One Man's Justice*, 43.

253. *Trade Union Act v. Local 97*, 170–171.

254. Ibid., 175.

255. Berger, *One Man's Justice*, 44.

256. *Trade Union Act v. Local 97*, 179.

257. Ibid., 180.

258. Ibid., 184–185.

259. Ibid., 179–181.

260. Berger, *One Man's Justice*, 47.

261. Ibid., 48.

262. *Trade Union Act v. Local 97,* 209–210.

263. Ibid., 235.

264. Ibid., 237.

265. Berger, *One Man's Justice*, 49.

266. Ibid., 50.

267. *Trade Union Act v. Local 97*, 250.

268. Ibid., 251–253.

269. Ibid., 292.

270. Ibid., 333.

271. Ibid., 340–341.

272. Anonymous letter, *Trade Union Act v. Local 97*, 407.

273. Berger, *One Man's Justice*, 59.

274. Dan Illingworth, *Vancouver Province*, Jun. 7, 1960.

275. Ironworkers Local 97, *Conquering the Challenge* (Lynn Communications Group), 23.

276. *Victoria Colonist*, Nov. 21, 1959.

277. *The Province*, Dec. 2, 1959.

278. *Nanaimo Free Press*, Dec. 29, 1959.

279. *Vancouver Sun*, Feb. 6, 1960.

280. *Victoria Times*, Mar. 18, 1960.

281. Bud Elsie, *Vancouver Sun*, Jan. 21, 1960.

282. Ibid.

283. John Holme,*Vancouver Province*, Jun. 7, 1960.

284. John Olding, *Vancouver Province*, Jun. 7, 1960.

285. John Arnett, *Vancouver Sun*, Jun. 8, 1960.

286. Len Norris, *Vancouver Sun*, Jun. 8, 1960.

287. Paddy Sherman, *Vancouver Province*, Jul. 14, 1960.

288. Jack Lee, *Vancouver Sun*, Jul. 7, 1958.

289. Reverend R.R. Cunningham, uncredited newspaper clipping (Jun. 20, 1958) from North Vancouver Museum & Archives.

290. Bob Harris letter to James Morris (friend of McDonald family), Jan. 19, 1996.

291. Uncredited newspaper clipping (Mar. 27, 1954) from North Vancouver Museum & Archives.

292. Doug Ward, *Vancouver Sun*, Mar. 10, 2007.

293. R.S. Eadie, Royal Commission, 682.

Index